Carpet Monsters and Killer Spores

CARPET MONSTERS AND KILLER SPORES

A NATURAL HISTORY OF TOXIC MOLD

Nicholas P. Money

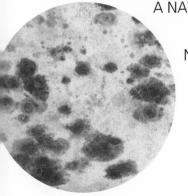

OXFORD
UNIVERSITY PRESS

2004

OXFORD
UNIVERSITY PRESS

Oxford   New York
Auckland   Bangkok   Buenos Aires   Cape Town   Chennai
Dar es Salaam   Delhi   Hong Kong   Istanbul   Karachi   Kolkata
Kuala Lumpur   Madrid   Melbourne   Mexico City   Mumbai   Nairobi
São Paulo   Shanghai   Taipei   Tokyo   Toronto

Copyright © 2004 by Oxford University Press, Inc.

Published by Oxford University Press, Inc.
198 Madison Avenue, New York, New York 10016

www.oup.com

Oxford is a registered trademark of Oxford University Press

Library of Congress Cataloging-in-Publication Data
Money, Nicholas P.
  Carpet Monsters and killer spores : a natural history of toxic mold / Nicholas P.
Money.
    p.   cm.
  ISBN 0-19-517227-2
  1. Molds (Fungi)—Control. 2. Molds (Fungi)—Health aspects. 3. Indoor air
pollution. 4. Dampness in buildings. 5. Dwellings—Maintenance and
repair. I. Title.
TH9035.M65 2004
648'.7—dc22   2003064709

9 8 7 6 5 4 3 2 1

Printed in the United States of America
on acid-free paper

For Allison, my stepdaughter

Preface

My colleague Jerry McClure was featured in the preface to my first book, *Mr. Bloomfield's Orchard*, but I didn't expect he'd make his way into this one. Jerry has a knack, however, for saying the right thing at the right time. Passing me in the hallway outside my lab last year, he greeted me by saying, "You're in the wrong business." This was a little unsettling, because I'd always thought of him as a supporter. Then he explained, "You could be making millions from black mold." I didn't think much about this pearl of Texas wisdom at the time, but it festered in my subconscious until my publisher asked me if I knew anything about indoor molds. Bowing to the lobbyists, I put aside my research for a bestseller on the organisms that squirm in the foul pond in my backyard and set off in search of indoor molds.

Few Americans can be unaware of the toxic mold crisis and the crisis of toxic mold lawsuits. The ravages of the mold *Stachybotrys*, and the ensuing legal battles between residents of sick houses, parents with sick children, building contractors, landlords, and insurance companies, are regularly showcased in newspapers and on television programs. As a mycologist, I had read about cases of mold-related illness long before Jerry McClure's interjection, and had even earned a few car payments by consulting on mold problems, but none of this had captured my interest until a black mold attacked my wife.

I had bought Diana a gift box of hand lotion, soap, and lip balm that trumpeted an all-natural, no-preservative pedigree. She loved the lip balm and took the little jar of the stuff with us to Norway in the summer of 2002 and used it every day. And then her head exploded in the bathroom, blackening the mirror with soot— not quite. First, her lips tingled; then, a few days later, they became inflamed every time she applied the gel. She stopped using the balm but packed it back to Ohio. Getting ready for work one morning, she opened the jar and was alarmed by the discovery that the balm had turned jet black. Being a science nerd, she guessed what had happened. A microscopic mold was eating the fat molecules in the cosmetic.

Looking at the black stuff under the microscope, the identity of the fungus was clear: *Aspergillus niger*. This is a common microorganism that forms masses of spores at the tips of stalks. Each stalk looks like a brush, which gives the fungus its name: An aspergillum is a special brush used to sprinkle holy water. *Aspergillus niger* can cause serious infections, particularly in patients whose immune defenses have been compromised by viral infections such as HIV, by cancer therapy, or by the anti-rejection drug regimens after an organ transplant. A few strains of this mold also produce toxins, but Diana survived the encounter with nothing worse than the knowledge that she had been smearing allergenic spores on her lips.

Like the media darling *Stachybotrys*, *Aspergillus niger* qualifies as a black mold because it is a microscopic fungus whose spores are painted with melanin. There are a few thousand species of these fungi and, in all likelihood, millions of distinct varieties or strains, each with their own genetic and biochemical character. They consume almost every kind of substance produced by plants and animals, which, if you think about it, includes most of the stuff we eat and many of the things used to construct a building. The diet of molds evolved over hundreds of millions of years, and far from evicting fungi, humans have only succeeded in broadening the range of food materials available for fungi by developing crops, transforming trees into paper products, and creating synthetic materials such as plastics, lubricants in machines, and fabrics. Any attempt to completely oust fungi from the indoor environment, or from any other place, is futile.

Negative consequences of interactions between molds and humans—as well as all of the positive ones—have been recognized for a long time. For instance, *Stachybotrys* was first identified in 1837. But widespread fears about black-mold toxins are a product of the new millennium and deserve a critical, balanced, scientific inquiry. Though I cannot promise anything that boring, I do hope I can dispel some of the media myths about these microorganisms while identifying the real threat that can be posed by a few of these fungi. Beyond black molds, other fungi are lurking in our basements, species that may someday replace *Stachybotrys* & Company as the new menace, a cash cow for the legal profession and a bane of insurers. These beasts will be featured in the final chapter.

The target audience for this book includes three groups of potential readers. First and foremost, I hope that anyone interested in the safety of their homes will find this book helpful in understanding the science behind the hysteria. Those in the legal profession and insurance industry are a second audience. Whether you are arguing for or against those who claim to have been victimized by mold, you will find some useful quotes. Finally, as a scientist, I enjoy writing for other scientists and shall use this book project to further explore the biology of the fungi.

For the reader(s) who enjoyed *Mr. Bloomfield's Orchard*, I should mention that I'm back in my writing shed, a couple of years older, definitely no wiser. This is a genuine preface, in the sense that it was written in advance of any of the chapters. This done, all that remains is to indulge in serious caffeine abuse for a few months and get the rest into my computer before the molds, or their attorneys, get to me.

Now that the book is written, I wish to thank everyone who responded to my barrage of e-mails and phone calls. The following people deserve special mention for indulging my queries: Luis De La Cruz, Dorr Dearborn, Richard Haugland, Steve Moss, Ian Ross, Steve Vesper, and Mike Vincent. I also thank my editors Diana Davis, Kirk Jensen, and Niko Pfund.

Nicholas P. Money
Oxford, Ohio
October 2003

Contents

Carpet Monsters and Killer Spores

If you have found where you were

exposed to toxic mold, you will need

to find the responsible parties (if any)

who are capable of paying.

—http://www.toxic-mold-stachybotrys.com (2002)

Stachybotrys versus Superpower

I am deeply in love with fungi, enough to cloister myself in a backyard shed for an entire year, through freeze and broil, to publicize their charms in *Mr. Bloomfield's Orchard*. So just as someone must be very bad if their mother says so, my opening statement should carry some weight. Some molds can be our enemies. Black-colored ones have been in the news a lot lately. They invade our homes, provoke allergies in millions of people, are implicated in a lethal respiratory illness in babies, and are a potential source of biological warfare agents. This is their story. Go make some tea. This will take a while.

Once upon a time, hundreds of millions of years ago, microscopic fungi figured out how to decompose plant and animal tissues. Hundreds of millions of years later, *Homo sapiens* evolved and learned how to construct shelters from leaves, branches, and animal hides. The fungi grew on the first midden assembled by very proud and very hairy Adam and Eve, but this didn't concern the newlyweds because the roofing and walls were easily replaced once they began to rot. The rare asthmatic child in those days was severely challenged by this intimate association with the fungi, but with

3

thousands of years to go before the invention of bronchodilating inhalers, the little chap wheezed himself to death before he could reproduce, preventing the spread of his asthmatic genes. When we left the forests, the molds followed in clouds of spores and rode on the wood that we dragged onto the plains. They have never gone away. To these microorganisms, a Rhode Island mansion is just as tasty as a hut in Zimbabwe. During their lengthy history, black molds and other fungi have perfected a method for transforming dense plant tissues into syrup. This is an impressive trick, because it allows them to feed on a multitude of food sources in nature, and also enables them to thrive on plant products inside homes, including wallpaper, paper-wrapped drywall, and particle board. All of these materials are made from plant fiber or cellulose. When we add the astonishing spore-producing potential of the molds to this picture of biochemical virtuosity, it is evident that we are in conflict with an invincible group of organisms. While there are solutions to mold damage in homes, humans cannot triumph over fungi in any wider engagement. Molds are here to stay.

The Internet offers a fine introduction to the contemporary mold problem. A search for "black mold" using any of the more effective search engines identifies more than half a million sites that feature these fungi, and "toxic mold" generates more than 100,000 hits.[1] "Toxic Black Mold Lawsuits" proclaims one site (www.toxic -mold-stachybotrys.com), offering referral to lawyers for those whose health or property have been damaged by a fungus. The banner at the head of the page shows a photograph of some spores, and a microscope and stethoscope that might have come from a child's science set. I hope to convince you that our understanding of molds and mold-related illnesses probes a little deeper than the intelligence furnished with a plastic microscope. But from a legal standpoint this site is quite good. Visitors are presented with a number of caveats, including "If you have been injured by a mold, that alone is not grounds for damages." (The grammar needs some tweaking.) News about mold lawsuits is also furnished at www. blackmoldclaims.com, a site that sports a "blackMold" headline whose letters drip down the page like wet paint. This site also provides good advice, but other entrepreneurs eliminate caution and offer endless wealth for everyone fortunate enough to have been attacked by mold. Black molds have become celebrity microbes.

Hundreds of different species of microscopic molds with black spores have been catalogued by mycologists, but *Stachybotrys char-tarum* (pronounced *stack-ee-bot-riss chart-are-rum*), also known as *Stachybotrys atra*, is the only one with wide name recognition (figure 1.1). In popular usage, the Latin name *Stachybotrys*—which means "grapes on a stick"—has become synonymous with black mold (*chartarum* refers to its growth on paper and *atra* means black). The infamy of this particular fungus is due to its production of poison-ous compounds called mycotoxins. But other molds are usually more prevalent in buildings, and even when *Stachybotrys* appears, the presence of the toxins is not guaranteed. In the current climate of rampant litigation, these subtleties are often ignored. A scattering of newspaper articles on indoor molds appeared in the early 1990s, and when concern about mycotoxins grew at the end of the millen-nium some observers dismissed this as another example of the wider malaise christened Y2K. But these were early days for the mold crisis, and journalists found more time for *Stachybotrys* once Y2K had passed and everyone had forgotten about the bunker-bound lunatics of December 1999. Media stories and lawsuits involving molds entered a logarithmic phase of multiplication, and insurance companies became overwhelmed with mold claims made by hom-eowners. A single insurance company in Texas handled 12 cases in 1999. The next year, the number increased to 500, and in 2001, the company fielded more than 10,000 claims.[2] Many of the insurance claims led to lawsuits: The firm was confronted with an average of 30 to 40 new lawsuits per week in Texas in 2001.

The appearance of extensive mold growth in a home is bound to be a demoralizing experience (plate 1). Nobody is going to feel

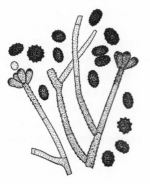

Figure 1.1
Spore-producing stalks (conidiophores) and black conidia of *Stachybotrys char-tarum*. (From M. B. Ellis, *Dematiaceous Hyphomycetes*, Wallingford, United Kingdom: Commonwealth Mycological Institute, 1971. Reprinted with permission.)

happy if their walls develop hundreds or thousands of black bruises. But anxiety really sets in when the occupant learns that the disfiguring bloom may be toxic: This piece of information turns a sanctuary into a prison. Photographs don't do justice to the horror of a building defiled by fungi. Remember, I adore fungi when they grow in the woods and in the laboratory, but while researching this book I often felt an overwhelming sense of repulsion when I witnessed their exuberance in a home. Here's an example. Home restoration expert Jim Moss took me to a bungalow—by then abandoned by the owners—that was past saving. Nastiness wafted though my face mask into my nostrils when we opened the front door, and the air was foggy with spores. The stench of decay produced by indoor molds is difficult to describe, but might (I'm guessing) bear some similarity to a sumo wrestler's laundry basket. The walls and ceiling presented tapestries of multicolored colonies, and with each step my shoes became soiled with spores from the festering carpets. This seemed more fungus than house. After a few flashes from my camera, we stepped back into the sunshine, greatly relieved by our escape from the mold banquet. This isn't a trivial problem. Children had slept in those spore-spattered bedrooms.

Stories about celebrity victims attacked by the celebrity microbe have done a lot to boost public awareness of the problem. Ed McMahon, Johnny Carson's affable sidekick on *The Tonight Show* for 117 years, was pursued into retirement by a black mold that festooned the walls of his dwelling in Beverly Hills. Ed and his wife developed coughs and migraine headaches, and their dog, sensitively described by the *Los Angeles Times* as "a mutt called Muffin that resembled a sheepdog,"[3] suffered from a severe respiratory illness and was put to sleep. *People* magazine published a photograph of Ed and his dog with the caption reading "McMahon blames mold for the death of Muffin." Articles about the mycological assault on Ed's residence mentioned that members of the household staff were also sickened, but any butlers, chauffeurs, cooks, maids, stewards, or valets[4] took the back seat to Muffin's chesty cough in the newspapers. The McMahons filed a $20 million lawsuit against their insurers and the testing and remediation companies hired to combat the mold.[5] The plaintiffs claimed that the problem began when a contractor failed to fix a ruptured pipe that flooded the den. Worse still, they had been advised to stay in the home during a botched cleanup, as black mold crept into their bedroom through ductwork and impregnated their clothes. Erin Brockovich, who became famous for winning a $333 million judgment against a company that had poisoned the water supply of a town in the Mojave Desert (and whose breasts were played by Julia Roberts in a movie of her life story), is another well-known casualty. Her home near Los Angeles, purchased with movie royalties and the bonus from her legal victory, was seriously violated by mold and necessitated a $600,000 cleanup operation. Later, while testifying against *Stachybotrys* before the California Senate Committee on Health and Human Services, she declared, "I wasn't looking for mold—mold found me." A similar sentiment was expressed by Melinda Ballard, a Texan whose experience with mold damage, and the resulting $32 million legal judgment against her insurance company, turned her into an instant celebrity. I'll reserve her story for a later chapter.

Unfortunately, fungal spores did not evolve as a benefaction intended to remind us that vaults full of banknotes and jewels do not ensure happiness; instead, indoor molds illustrate Karl Marx's contention that the proletariat bears most of the burden. I live in southwestern Ohio, between Cincinnati and Dayton, so I'll furnish a couple of stories of mold attack from my area that have led to lawsuits. (The location is, however, of little significance. Every city newspaper in the country offers a mine of information and disinformation about indoor molds.) Molds evicted Sheila Marshall from her home in West Chester, an affluent township on the northern edge of Cincinnati favored by commuters. Her encounter with fungi, featured in a series of articles in the *Cincinnati Enquirer* written by journalist Michael Clark, serves as a typical case. Sheila's home became seriously contaminated with black mold soon after construction, and she began suffering from fatigue, memory loss, dizziness, and allergies. A "mold expert" confirmed that spores of different fungi were present, including low levels of *Stachybotrys*. Clark informed readers that "All are considered potentially deadly, especially to those with compromised immune systems."[6] Sheila left her home on the advice of her physician, who diagnosed her symptoms as "toxic mold syndrome." Poor construction and persistent water leaks were blamed for the fungal contamination, and the builder was challenged with a $75 million lawsuit for failing to correct the construction defects. This was a devastating experience, both for Sheila and her builder.

Another example will help flesh out this portrait of the common kinds of mold problem reported across the country. The *Cincinnati Enquirer* also publicized the trials of the Vanden Bosch family, who were unaware of mold growth in their house until they began removing wallpaper for a home improvement project.[7] Mold had burgeoned on the drywall under the paper and the homeowners were horrified when a contractor found the stuff under carpets and blanketing wooden subflooring. Peter Vanden Bosch, 8-year-old son of Tom and Mary, had suffered from upper respiratory infections and headaches for some time, but these symptoms worsened following the removal of the wallpaper. Mary developed a severe rash and was "covered in hives." *Stachybotrys* and a second mold called *Aspergillus* were identified in the Vanden Bosch's home. After the contamination was discovered, Mary was tested for a potentially cancerous skin lesion, and to determine whether "the mold and its toxins [were] in her bloodstream." No reputable scientists have countenanced a link between cancer and indoor mold exposure, but the newspaper article dispensed with this obstacle to drama, stating that molds are "linked to numerous health problems, including some forms of cancer, skin irritation, systemic infections, hemorrhage and convulsions." Understandably, the family became very alarmed when a contractor donned a biohazard suit and respirator before he entered their home. This convinced them to move to a hotel while the infestation was treated. Their insurance company was held liable for most of the $50,000 spent in cleaning the house and replacing damaged drywall and carpet.

In addition to stories about individual families afflicted by mold damage, news reports have covered housing developments in which multiple homes have become riddled with fungi, and toxic apartment complexes saturated with mold spores. Reviewing newspaper articles and television news stories concerned with molds, I have been struck by the attention given to homes in prosperous neighborhoods. Do molds avoid the housing projects of Cincinnati and Dayton? Perhaps the buildings in the projects are constructed with greater care than houses in the commuter belts? But when a home is torched because its mold contamination is beyond control, local news teams assemble for the bonfire without regard to the original value of the kindling. The reportage is worse than formulaic, and invariably includes a journalist goading a miserable homeowner or renter with a microphone to provoke an answer to that most asinine question: "How do you feel?"

Besides homes, other buildings including hotels, manufacturing plants, and schools are prey to fungi. Teachers from the high school and middle school in West Carrollton, a suburb of Dayton, filed a $6 million lawsuit against their school district claiming that they had been sickened by mold exposure and could no longer teach. They described a range of symptoms including sinus infections, headaches, memory loss, and an inability to concentrate.[8] School district officials denied the claims made by the staff, partly because there were no initial reports of mold-related illness among the students. The story thickened, however, when students began complaining of illnesses connected with the mold damage, and staged a walkout to protest the condition of the school buildings. Schools throughout the area have suffered mold damage, and millions of dollars have been invested in purging fungi from classrooms and in preventing their return.

It is not an exaggeration to cast the mold problem in Ohio and the rest of the country as an epidemic. But to fully comprehend the seriousness of indoor mold growth, and its danger in individual cases, it is necessary to explore the toxic mold story in greater depth than any newspaper article. What is the extent of the fungal damage in the home? Is there any clinical relationship between the appearance of the fungus and a homeowner's illnesses? If so, what toxins, or allergenic compounds, are responsible for the symptoms? And, to return to the opening quote for this chapter, who (if anyone) should be held responsible for eradicating the fungi? All of these questions will be addressed in this book.

Contrary to statements made by some commentators on the mold epidemic, the effects of mold exposure on human health are very hazy. A disease called stachybotryotoxicosis was identified in Ukraine in the 1930s, where horses and other animals fed on straw contaminated with *Stachybotrys* developed large skin bruises, suffered massive bleeding of the intestine and other organs, and displayed nervous disorders. The human version of the ailment appeared in people who handled infested straw or slept on straw-filled mattresses. Patients developed dermatitis, inflammation of the mouth and throat, and suffered nose bleeds, fever, and headaches. Studies by Soviet scientists concluded that compounds called trichothecenes (*try-coe-thee-seens*) produced by the mold were responsible for outbreaks of stachybotryotoxicosis on collective farms.

The first report of trichothecene poisoning caused by molds

growing in homes appeared in a paper published in 1986 by William Croft and colleagues.[9] The study was concerned with a single, mold-infested home in Chicago in which the whole family developed flu-like symptoms, diarrhea, headaches, fatigue, and dermatitis, and suffered hair loss. Numerous roof and plumbing leaks in the home fostered extensive mold damage, and air sampling revealed high concentrations of *Stachybotrys* spores. The interior of a heating duct was coated with an inch-deep felt of spores mixed with lint and carpet fibers. Drawing on various lines of evidence, Croft pointed to an association between the mold and the household's medical problems. When samples of the black gunk collected from contaminated areas were injected into rats and mice, the animals died within 24 hours. Histological study of the rodents revealed hemorrhaging of blood vessels in major organs, which seemed consistent with the effects of trichothecene poisoning. Finally, chemical analysis of contaminated fiberboard collected from the home identified several potent toxins characteristic of *Stachybotrys*. At the time, the study did not attract a lot of attention. It served as another example of the diverse relationships between humans and fungi, and was viewed as an isolated event of minimal concern to clinicians and public health officials. That picture changed between January 1993 and December 1994, when 11 Cleveland babies were hospitalized with bleeding lungs.[10] Parents brought their infants to hospital when they developed nosebleeds or began coughing up blood. Dr. Dorr Dearborn, at the Rainbow Babies and Childrens Hospital in the city, diagnosed the condition as idiopathic pulmonary hemosiderosis (IPH). "Idiopathic" means that the cause of the illness is unknown; "hemosiderosis" refers to the accumulation of iron inside the lung— the iron comes from the red blood cells that spill into the lungs when the surrounding blood vessels begin to leak. Some of the infants suffered repeated episodes of lung bleeding after they were treated and returned home. One 10-week-old boy died of respiratory failure.

Normally, pediatric pulmonary hemorrhage is a very rare condition, occurring in about one in a million babies. In the decade preceding the cluster of cases in Cleveland, lung specialists at Rainbow had treated only three infants for the illness. Recognizing that something unusual was happening, Dearborn alerted the Centers for Disease Control and Prevention (CDC) in Atlanta, and a case-control study was initiated to identify common factors that may have predisposed the babies to lung damage.[11] The initial study concerned 10 patients ranging in age from 6 weeks to 6 months; nine were boys, all were black, and all lived in the eastern part of the city. Dearborn and colleagues began by comparing the medical records of the sick babies (cases) with those of 30 healthy babies (controls) selected at random from the same zip codes. Race was an obvious consideration, but the researchers recognized that the correspondence between skin color and economic status was more likely to be important than any genetic propensity to disease. The medical history of the infants didn't offer any insight, because the babies had been in excellent health before they stopped crying, became pale and limp, and blood oozed from their mouths and noses. This led the investigators to examine the patients' homes. During the summer of 1994, Cleveland experienced some of the heaviest rainfall in its history. Flooding was reported in the eastern part of the city, and in the month before the lung-bleeding episodes all of the case homes had been water-damaged. The investigators found that there had been little or no effort to clean flooded areas in the homes, and in some instances the mess was exacerbated by leaking roofs and faulty plumbing.

As the study progressed, the investigators found that molds had proliferated in the homes of the sick babies, and that *Stachybotrys* spores were abundant. Because *Stachybotrys* was known as a source of trichothecene toxins that were associated with the hemorrhaging of blood vessels in animals, the CDC tentatively identified the black mold as the cause of the bleeding lungs. Once the mold connection was established, additional cases of lung bleeding in Cleveland were reexamined, and Dearborn and the CDC uncovered others whose symptoms and home environments were comparable with the original group. Two of these babies had died. They then trawled through the coroner's records and reexamined every infant death in Cleveland between January 1993 and December 1995. Of 172 recorded deaths, 117 were attributed to sudden infant death syndrome, or SIDS. (The cause of SIDS has been studied, without resolution, for more than 200 years.) Specimens of lung tissue obtained during the autopsies of the Cleveland SIDS babies were obtained from storage, and it appeared that six of them had bled into their lungs and might have suffered from the same symptoms as the victims of IPH treated at Rainbow. Could black mold have been the cause of all this misery? Was *Stachybotrys* a baby killer whose activities had been masked by the confusion surrounding SIDS? Physicians in other parts of the country began reporting patients with similar symptoms. Suddenly, it seemed that toxic fungi were everywhere.

Concern about indoor air quality (IAQ) developed in the 1970s, when increasing numbers of office workers complained of headaches, dizziness, nausea, and other ailments. Inadequate ventilation and chemical pollutants were identified as likely contributors to these vaguely defined illnesses. But in the absence of a specific cause, the term "sick building syndrome" was coined as a catchall diagnosis for most patients. The related term "building-related illness" was used for respiratory disease and other conditions whose cause could be identified. Legionnaires' disease and Pontiac fever caused by the bacterium *Legionella* are examples of building-related illness. Indoor mold spores were mentioned in relation to allergies, but fungi weren't viewed as a major contributors to IAQ problems—at least by the public. The outbreak of lung bleeding in Cleveland changed this. By the late 1990s, the media had elevated ubiquitous molds to the status of life-threatening microorganisms whose appearance transformed homes, schools, and workplaces into hazardous environments. Buildings needed to be tested, and toxic ones needed to be cleaned. These tasks were embraced by industrial hygienists who had dealt with IAQ problems before molds hit the headlines, and new job titles were printed on business cards: mold inspector, mold contractor, and mold remediator.[12] A new industry was born.

The first task for the mold inspector/contractor/remediator brought into a suspect building is to determine the severity of the contamination. Because mold spores, or fragments of mold colonies, are responsible for allergic reactions and may serve as airborne vehicles for toxins, information on the concentration of these microscopic particles seems crucial. Spore concentrations can be measured with a sampling device that draws a known volume of air through an inlet and deposits particles on a microscope slide or culture dish. The number of spores on the glass slide can be counted directly. This slide count does not discriminate between living spores and dead ones, but, since the dead ones can still cause an allergic reaction and carry toxins, this subtlety is unimportant. The number of colonies that grow out over the agar in a culture dish usually provides a lower estimate of mold concentration, because the method reflects the number of live spores capable of germinating—not the total number of spores. Different authorities have set limits for indoor spore concentration between 50 and 500 live spores per cubic meter of air, but variations in individual sensitivity to inhaled allergens, and possibly to any toxins carried by the spores, render these numbers of dubious value. Some people are untroubled by rooms filled with mold spores, while others seem debilitated by exposure to very modest concentrations.

Unless a home is severely contaminated with mold colonies, the spores swirling indoors come from outdoors. Mold reports and forecasts in newspapers and on television are based on data that is updated every day for major cities in the United States.[13] Spore counts across the nation average above 1,000 per cubic meter but can drop to zero during dry weather and run to tens of thousands after spring downpours. Most of the spores in outdoor air come from fungi growing on plants. When windows are open, the mixture of spores in a room soon matches the composition in outdoor air. Comparisons between the indoor and outdoor spore counts can be helpful in evaluating the level of home contamination, but only if the number of spores of each species are tallied. *Stachybotrys* spores might be common in a damp bedroom, for example, but rare in samples taken in the yard, even though total spore counts are the same. The contractor who reports that the spore concentration in a home exceeded 2,500 per cubic meter must provide a great deal of additional information to allow the homeowner, and the courts, to make sense of this information: Where were the measurements made? How were the measurements made? What types of spore were found?

Despite their perceived importance, however, spore counts are next to useless for assessing many indoor mold problems. The spores of some molds, including those of *Stachybotrys*, are sticky cells that become bonded to the surface of the colony as they dry. This means that they do not drift around unless their substrate—the material on which they are growing—is disturbed. An unscrupulous contractor can elevate the number of airborne spores in an indoor environmental survey simply by banging on a wall or contaminated air conditioner before collecting an air sample. Obviously, spore concentrations in air are greatly affected by the circumstances at the time the sampling is performed, either by a pair of lungs or an automated spore sampler. More effective analysis of spores attached to surfaces or settled as dust entails scraping small samples from contaminated materials into a plastic bag, using a vacuum sampler, or removing spores from a surface with a strip of clear adhesive tape (the tape-lift method). The identity of the fungi can then be determined by microscopic examination. If you have followed these details of the exhilarating life of the industrial hygienist, you may have posed the following question: Spores that aren't swirling around in the air can't cause any problems, so why bother about them? The reason that molds on surfaces should be studied is because they always represent a potential source of airborne spores.[14]

Ignoring measurements of the number of spores in air or on surfaces, there is an unscientific "f*** me!" response to a mold-contaminated room that is very dependable when it is provoked in someone who has peered into a lot of wet buildings. There has been an attempt to endorse this by establishing threshold levels for fungal growth that warrant different reactions. Less than 10 square feet of mold damage, clean the area yourself; 10 to 32 square feet, time to get more serious and remove contaminated materials.[15] If more than 32 square feet of wall are covered by mold growth—by which I mean sufficiently defaced to provoke involuntary expletives—then one should consider calling remediation or abatement specialists. Similarly, if your kitchen ceiling sports a mural of star-gobbling black holes, it's time to pick up the phone. Which brings me to remediation methods.

The eradication of small-scale fungal growth is, or at least it was once, a housekeeping chore. Mold can be cleaned by scrubbing the contaminated surface with diluted bleach or detergent (a toothbrush works wonders for mold growing around faucets or other plumbing fittings). The CDC recommends 1 cup of bleach in 1 gallon of water. If the bleach makes contact with the cells of the fungus, they will be destroyed, but the problem is more complex if the fungus is growing on both sides of the paper wrap on drywall or has permeated sections of particleboard. In these situations, the contaminated material should be discarded and the area repaired. Synthetic biocides containing ammonium ions, hydrogen peroxide, copper, silver, or tin[16] can be very effective at killing indoor molds, but only if they saturate the target. Spores can also be removed from nonporous surfaces by vacuuming. As the size of the damaged area increases, efforts must be made to contain the fungus during the cleanup to limit further spread of the spores. This can be done with polyethylene sheeting and duct tape. Protective clothing is probably justified for those involved in the cleanup. You might think that a mold contractor who wears a respirator leads a dangerous life, but the most frequent job-related accident in this profession occurs when workers splash themselves with concentrated bleach. Whatever efforts are made to clean contaminated areas and remove damaged materials, they will be futile unless the source of the moisture is located and fixed. If the water keeps coming, the fungus will return within weeks or even days.

Even without a leaking roof or pipes, a humid indoor climate can promote fungal growth. The ceiling above a shower isn't usually splashed with water after a few years of marriage, but the water condensing from steam onto the painted surface is often enough to slake the thirst of a mold colony. If the bathroom suffers from poor air circulation, the fungi will be even happier. Give me a few minutes in any McMansion and I'll bet I can find mold growth in its multiple bathrooms. Nobody lives mold free, but appreciable mold growth cannot occur without plenty of water. Industrial hygienists like to measure indoor moisture levels, because, though this is simpler than making a sandwich, the use of a handheld meter imparts an air of brilliance that will persuade a hapless homeowner that their security is assured by the hands of a master. Humidity measurements can be useful from different areas of a home because the moisture level in a loft or crawl space can reach saturation without having any obvious effect upon the interior of a room. The water content of solid materials like drywall can be measured with a meter fitted with sharp spikes that are pushed through the painted exterior. These measurements are more helpful than humidity estimates from the air, because they indicate the potential for mold growth in specific locations. On an even finer scale, the water content of a single wooden beam can range from bone dry along most of its length to sopping wet around a nail. These "microenvironmental" problems are usually beyond detection. Wet spots are prime sites for mold growth, but the more obvious test for mold is to look for the stuff itself. Moisture measurements are an overrated tool, but they are important in the courtroom because the jury will not be impressed if the expert witness answers "really wet" or "my glasses steamed up" when questioned about the dampness of a property. Advice to the expert witness: Give the jury numbers, and defend your numbers (rather than your car payments) as if your life depended upon them.

Unless a mold problem is likely to lead to a lawsuit, I'm not convinced that anyone should pay a contractor to collect air samples and make moisture measurements. Recently, I was asked to look at overt fungal damage in a house that was being sold to a renter. The person complained of terrible allergies whenever she slept in the house, so I cannot imagine why she wanted to buy it. The property had faulty drainage that regularly soaked the basement. One of the bathrooms suffered from an extensive growth of mold. The wall behind the mirror and bathroom cabinet, and every corner and seam that collected water, were blackened with spores. The rustiness of screw heads and other metal fittings also indicated a humidity problem. I brushed some of the blackness into a plastic bag and soon came up with a list of the fungi that were growing in the home: *Aspergillus*, *Cladosporium*, and *Alternaria*, but no *Stachybotrys*. The drainage in the backyard had to be fixed by diverting rainwater away from the foundations, and the home needed the services of a professional cleaning company. In the meantime, I suggested that the windows should be opened whenever possible to allow air to circulate freely and help dry the walls. This is typical of the majority of cases of mold growth repeated in every town in America. It didn't warrant hours of testing and a report in a fancy binder, there was nobody that the renter could sue, and my involvement did nothing to help me buy a Ferrari.

A wet building isn't necessarily a moldy one. Stimulated by recent news coverage, the occupants of another building I examined were convinced I'd find something fungal and probably toxic. As soon as I opened the door to the stairs, I recognized the smell: old English church. Damp and musty, but no hint of a sumo wrestler's *mawashi*. Water was seeping through the cinder block wall, leaving a three-story high column of damp patches. Some areas of the wall were covered with masses of white fluff which was sloughing from the surface and collecting on the stairs. What do you think? Fungal? A few days before my visit, the building had been inspected by an industrial hygienist who had collected samples and sent them to "an American Industrial Hygiene Association accredited laboratory" for an "anion scan" and determination of their calcium, potassium, and sodium content (dissolved in water, these three elements are positively charged cations). The chemists concluded that the samples "contained significant chloride . . . and sodium," and from this, deduced that water damage might have contributed to the development of the white stuff (without saying what the white stuff was).

Think of the stupidest person you have ever met. Go on, indulge me. (I remember a girl in my elementary school who said, "It has eight legs, Miss," when shown a drawing of a horse in which each limb was drawn as two descending parallel lines.) Now, if you were to lead your imbecile into a humid stairwell and point to the wet patches, don't you think he or she would verify that the building might have been water-damaged? Thank you, and here's a check for $5,000.

Looking at the white stuff with a hand lens, I felt sure it wasn't fungal. This was confirmed with a microscope: The white threads were crystals of calcium sulfate, or gypsum. (Why hadn't the element analysis detected significant calcium?) There was no sign of anything fungal. The building had no mold problem, just a leaking roof. Water was running on the outer surface of the wall, seeping into the building through the cinder blocks and carrying a slurry of dissolved calcium, sulfur, and other elements. Then, on the other side where water evaporated into the stairwell, the crystals grew as this slurry dried out. (This will make sense if you remember growing crystals in school—probably bright blue copper sulfate crystals—by allowing water to evaporate from a salt solution.) In this case, the investigator was well meaning, but this is no excuse. After all, we wouldn't accept an apology from a kindly physician who misdiagnosed dandruff as a brain tumor.

The rapid development of the mold scare has spawned a number of pseudoscientific approaches for eradicating fungi that will likely disappear when logic overtakes hysteria. Some companies have advocated the use of ozone for purifying the moldy air in homes. Ozone is a highly reactive molecule composed of three atoms of oxygen (O_3), rather than the usual pair of atoms that form the chemical carried by our red blood cells (O_2). It forms naturally in the upper atmosphere, creating that recently impoverished shield against damaging levels of ultraviolet light. Ozone is a very ironic molecule, because this chemical savior is also a lung-irritating chemical when it is created from automobile emissions. Astonishingly, however, this *pollutant* has been promoted as a treatment for asthma. To this end, ozone-generating air ionizers have been on the market for decades. My father bought one for me as I wheezed my way through an asthmatic childhood. The little box was placed next to my bed and made a buzzing sound as it electrified the air around the tip of a needle that projected from a hole in the front. The advertisement for this miracle device promised immediate results or money back; I continued to suffocate myself, so being a Money, my dad got his back. Like mold growth in a home, ozone has a smell that is immediately recognizable to the initiated: The same distinctive odor inhaled in smoggy cities is also whiffed far from traffic on beaches, where the stuff is created—through another natural process—from ocean spray. So sensitized am I by that childhood contraption that NASA should consider tethering me to a weather balloon and allowing me to report how far the ozone hole has shrunk.

The U.S. Environmental Protection Agency (EPA) has reviewed the available research on household ozone generators and concluded that ozone is ineffective at removing allergenic particles including mold spores and pollen from air, and that ozone may pose a health threat.[17] Even at low concentrations, it can cause shortness of breath, chest pain, and throat irritation. It may actually worsen the symptoms of asthma and compromise the immune response to respiratory infections. Yet sales of ozone generators remain strong, and television advertisements for the devices show healthy-looking models walking along pristine beaches breathing pristine air. In addition to creating ozone, some of the gadgets are supposed to trap airborne particulates by imparting an electrostatic charge to their surface that attracts them to a filter. Again, after evaluating these products, the EPA was profoundly unimpressed.

Some hygienists recommend using a scaled-up version of the bedside generator to pump ozone gas into contaminated homes at high concentration. Logic suggests that this must be next to useless for mold eradication. As the highly reactive molecules pass through a room, they probably kill spores in the air and on exposed surfaces, but unless the generator is kept on full blast, contaminated areas behind drywall and in confined spaces will escape. Mold cells called hyphae buried in the building materials are bound to be sheltered from ozone. The only thing that can be said with any certainty is that the treatment will damage rubber, wire coatings, fabrics, and artwork. Here's a final irony: The melanin pigment that blackens black molds functions as a protective coating that resists the damaging effects of ozone. Experiments show that compared with other microorganisms, black molds are especially well adapted to resist damage by ozone and other powerful oxidizing agents.

Another questionable approach to mold remediation involves freezing the fungus to death by spraying a contaminated wall with dry ice. The inventor of this ridiculous method calculated that he could make millions by offering to kill mold in every hotel room in America. Before investing in this technique, I suggest meditating on the possibility that super-cooling a wall would result in serious moisture buildup.[18] With so much money involved, there are stacks of pending patent applications thought up by crackpots all over the world. How about the certainty of a radioactive cure for mold? A 30-day no-obligation trial offer should convince anyone that the gamma rays streaming from a block of radioactive cobalt would kill every spore in a home. (Read the small print before investing: "The purchaser of this service holds responsibility for negotiating the evacuation of his or her neighborhood for a period of time not less than 35 years."[19])

The publicity attending mold problems has made identification of the various fungi useful, if only because homeowners' fears are soothed when they are told that certain molds are present, but that *Stachybotrys* is not among them. Not surprisingly, there have been instances of fraud in which consultants have fabricated reports by saying that a home was rife with *Stachybotrys*, simply to push up their fees. Other molds scams have been more inventive. A criminal ring in Texas purchased houses in the Houston area and then flooded them by cracking pipes and dousing the contents with garden hoses. After leaving the houses for a few days to allow mold to develop—a practice called "house cooking"—they reported mold and water damage to their insurance companies.[20] The ring collected $7 million from 54 false claims before they were arrested. Another scam in Houston involved a company called Mold Restoration, Inc., owned by Richard and Robert Steffan. Rather than cleaning homes, the brothers allegedly used insurance settlements paid to homeowners to pay for their gambling debts, club memberships, a yacht, a racing boat, new homes, motorcycles, and luxury cars.[21] The Steffans had limited mycological experience: Before their entry into the mold remediation business, they had operated a used-car dealership and had also marketed dolls, through a company called "Christians Are Us," that declared "Jesus Loves You" at the push of a button.

People manifest a range of responses to mold growth in their homes. Because we are in the early phases of the scientific investigation of illnesses related to indoor molds, it is difficult for anyone to measure the danger of mold exposure in a particular situation. Press coverage has certainly made many people terrified by the first hint of mold growth and has led to bogus cause-and-effect explanations for a diversity of ailments. Ignorance isn't bliss when it comes to fungi growing in homes, but there is no doubt that concern is overblown in many cases. I can think of one notable exception. A lifetime ago, when I was a mycology student in England, I visited an elderly gentleman who had telephoned saying that he had mushrooms in his bathroom and wanted someone to look at them. The sight and smell of the masses of brown cup fungi sprouting from the wood surrounding his bathtub made me grab the towel rail and clap a hand over my mouth (lovely cup fungi beneath the pine trees, but monstrous excrescences in a bathroom). The man brushed aside my response and said that he'd often looked at the cups as he soaked in the bath, and *over the years* had become intrigued by his bathmates. He wanted to know what they were and had no interest in getting rid of them. He was delighted when I told him that the cups were the fruiting bodies of an ascomycete fungus called *Peziza domiciliana*, which, as its name suggests, is common in homes.[22] The man had inhaled hundreds of millions of the spores that showered from the cups during his bathtime reveries but had not developed any symptoms of allergy. While few of us encounter masses of cup-fungus spores in our homes, there is no doubt that the spores of other fungi are part of the grime that pollutes the lungs of every human from first cry to last gasp.

Ecologists study organisms outdoors, in "nature," and view the environment within a modern home as the dullest of places. In comparison with the diversity of life that flourishes in the scrappiest of backyards—just look at the worms and beetles writhing in a pile of decomposing leaves—most homes are very sterile. Most of the biological action in a home takes place in the rich ecosystems harbored by the intestines of the human residents and their pets. But despite the spraying of disinfectants, sophistication of home plumbing, and ubiquity of antimicrobial agents in soaps, paper products, and even in clothes, plenty of things live with us. Even rubber-gloved disciples of Martha Stewart cannot sterilize their kitchens. The science of building microbiology has become a thriving field for researchers, not least, of course, due to our new fear of indoor molds. Every faucet and drain hosts a community of microorganisms; shortly after hanging, a shower curtain supports a greasy film of microbes called a biofilm. Molds flourish wherever something gets wet—and just about anything will do—and stays wet. Once a basement floods, a roof leak soaks a floor, or a broken pipe spurts into a wall cavity, indoor molds blossom on carpets and walls. Fungi do not spread themselves evenly throughout a house. A wet bathroom may be a veritable spore fountain, and though the fungi diffuse in a cloud when the bathroom door is open, most may fall to the carpet within a few feet. For this reason, an adjacent bedroom may be relatively clean, accounting for a persons' allergic response to a specific room or end of the house. Even in a single room, different types of fungi will be found on particular items of furniture, and the total number of spores will vary widely between chairs and carpeting.[23]

At the beginning of this chapter, I said that in many situations we are losing the battle against indoor molds. Failure is almost guaranteed when we have so little understanding of our adversary and have never stopped long enough to imagine what victory would look like. (I bet Alexander the Great or Napoleon said something similar.) What do we want? We can no sooner expunge molds from the planet than kill every other living thing, including ourselves. Fungi are an indelible part of life. Along with bacteria, and food webs of worms and other invertebrate animals, they play essential roles in waste disposal—biodegradation in the parlance of biologists. Indoor molds would be nothing more than an occasional nuisance without the assistance of shoddy construction and bad plumbing, and builders who build better homes for fungi than humans should be forced to correct their blunders or pay for the privilege of leaving the mess for someone else. Although litigation is essential to resolve many cases of mold contamination following water damage, the specter of a toxic mold crisis has also stimulated thousands of people with little to gripe about to join the class action lawsuit of every American versus every other American: "My child has a learning disability, and you built my moldy house, so give me all your money." There is nothing insightful in this view of our legal system, of course, but the toxic mold story offers a revealing example of the way in which the use and abuse of science can take center stage in the courtroom.

This book is offered as an escape from the melee—an island of logic, common sense, and calm. The following chapters will examine the biology of molds, explain how their spores cause allergies, and assess the evidence for their toxicity. Lawsuits by homeowners claiming to have been sickened by mold exposure have generated a remarkably acrimonious climate for those investigating the fungi

Stachybotrys versus Superpower

that grow in buildings, and my telling of the mold problem will also feature the battles between scientists, attorneys, insurance companies, and politicians. I'll also explain why and how fungi grow in homes, and introduce you to wood-rotting fungi capable of collapsing buildings. Unfortunately, all of this means that everyone needs to learn some mycology. Fortunately, you need only turn the page.

Is not the air we breathe charged

with them in the declining part of the year?

Do we not receive them into our Lungs

with every breath we draw?

—James Bolton, *Filices Britannicae* (1790)

Uninvited Guests

I assume that most of you were already aware that something important was happening in the electrifying world of molds before you picked up this book. Having introduced the controversy swirling around *Stachybotrys* and its toxins in the first chapter, I'm going to explain some of the relevant issues in fungal biology before offering an assessment of the toxic mold "epidemic." Besides its raw entertainment value, I promise this foray will be worthwhile. By the time you reach the middle of the book, your mycological expertise will allow you to speak the language of the mold inspector like a native, and to assess the sagacity of anyone offering to purify your poisonous home or office.

Most fungi are pigmented, and the majority of the pigments impart a duskiness to the walls of fungal cells that protects these otherwise diaphanous microbes from hazards in their environment. *Stachybotrys* owes its blackness to the production of melanins, similar to the pigments that darken human skin. Many other molds create more inspiring color schemes by weaving other types of pigment into their molecular architecture. Black, blue, green, or yellow molds dress themselves for life above the ground, where—as we know—temperatures fluctuate, lighting can be harsh, and water is often scarce. Why should a fungus leave the shelter offered by soil or water? Part of the answer is the abundance of plants that offer a seemingly limitless food supply for anything with the requisite digestive machinery. Soil fungi feed on plant tissues, too, but the fungi that can get into the air don't have to wait until a tree drops its leaves or topples over. Green plants use sunlight to manufacture sugar molecules, string them together in chains, and spin the chains into strands of cellulose that envelop the squishy interior of their cells. Cellulose is the most abundant polymer (big molecule) on the earth, and is a major food source for molds. This is the insoluble type of plant fiber that we must consume, but cannot digest. Many mold species defeat this physiological limitation. They possess the biochemical tools to fragment cellulose strings back into glucose, which explains how they can flourish on something as unappetizing as a damp kitchen cabinet.

To get at cellulose and other plant products, molds operate as pathogens, colonizing and sometimes killing living plants, and as saprobes, decomposing dead tissues. My backyard shed is a perfect example of a food source for molds. It was built a couple of years ago, and within months became blackened with spores. Only the flaky outer surface of wood has been colonized by fungi—perhaps encouraged by the nourishment in the purportedly mold-retardant wood stain[1]—so at their current speed they will be feasting on the shack for decades to come. Vegetarian molds need never go hungry. Besides wooden or wood-containing buildings, man-made mountains of cellulosic materials will feed fungi for millennia. Mount Rumpke is a 234-acre landfill outside Cincinnati that is visible from space. (No longer an impressive statement, since it has become possible to count someone's nose hairs using a camera on a satellite.) It is a mountain of waste paper and packaging. Billions of tons of mold fuel. Cellulose is certainly the best-selling entree for black molds, but the list of specials includes everything else manufactured by plants and animals: fats and oils, proteins, nucleic acids, and sundry molecules that keep cells alive or are minor parts of their substance. (Speaking of animals, I'm happy to forecast a brief discussion of one of my favorite topics later in this chapter: Fungi that dine on us.)

The term "mold" does not refer to a specific group of organisms, but is a commonly recognized description for any microscopic fungus.[2] The U.S. House of Representatives adds a few details in its proposed "Toxic Mold Safety and Protection Act of 2002" by defining mold as "any furry growth of minute fungi occurring in moist conditions."[3] This is very close to most dictionary definitions. Biologists regard all fungi as microorganisms, so how does teeniness enter the definition of a mold? Consider mushroom-forming fungi. Mushrooms are the conspicuous, aboveground manifestations of organisms that otherwise exist as masses of threads called hyphae within the soil. Because the hyphae that make up this subterranean feeding phase are microscopic, the fungus is inconspicuous, but it may cover a vast territory. The fungus becomes visible, and its expanse is revealed, when the hyphal threads commingle to produce mushrooms. Any fungus that lacks one of these conspicuous fruiting bodies is a mold. Like mold, "mildew" is used as a noun and an adjective and has an equally vague definition. Some types of plant infections caused by fungi are called mildews, and the fungi causing the diseases are types of mildew fungi; powdery mildews, for instance, are caused by powdery mildew fungi.[4] Mildew is also applied to mold growth that looks like plant mildew when it appears on damp paper, leather, or other products. A mildew, therefore, can also be described as a mold.

With this information in hand, I'll explain what a fungus is by telling you what a fungus does. Fungi are microorganisms that feed by absorbing nutrients with networks of hyphae and disperse their offspring in the form of microscopic spores. Hyphae are tubes of living cytoplasm separated into compartments in some fungi and existing as continuous conduits in others. The interior of the hypha is a whirr of fat globules and organelles that are carried along the tube on molecular conveyer belts. Some of the conveyer belts deliver new materials to the tip of the cell, and the cylindrical shape of the fungal hypha emerges from the resulting focused or polarized growth. Elongated human nerve cells and the pollen tubes of plants grow in a similar fashion, but do so for different reasons. Nerve cells elongate to create electrical circuits; pollen tubes extend themselves to deliver sperm cells to eggs inside flowers; hyphae function as microscopic mining devices. Hyphae probe, penetrate, and thoroughly permeate solid materials and extract nutrients in their path. As these cells extend, they pop out branches, and so give rise to a colony or mycelium that fans out like a river delta.

Let's consider a fungus growing in a toenail. Like hair, nails are made from keratin, which is a protein. Keratin is a tough material held together by the chemical bonds within its molecular structure. A toenail won't dissolve in water, at least not very quickly. But if the bonds within the proteins can be broken—and that's what some fungi do for a living—peptides and amino acids will be released from the chemical structure. These smaller constituents of proteins are soluble in water and are easily absorbed by the fungus to build fungal proteins or to fuel fungal metabolism. Hyphae release enzymes to cleave the bonds within the materials that they invade, imbibe the resulting molecules, and push themselves ever deeper into their food. The fungal way of life is a unique strategy for survival. Their evolutionary history has set them apart from the animal use of a mouth and a stomach, and the plant's solar-powered conversion of carbon dioxide into sugars.

Hyphal growth is a defining characteristic of the fungi, though some of them grow in the form of globular yeast cells rather than hyphae when they feed in soupy rather than solid environments. Mycologists recognize discrete groups of fungi. Fungi that form mushrooms and related plant parasites called rusts and smuts are called basidiomycetes. Morels and truffles are examples of ascomycete fungi, and bread molds are zygomycetes. Together with an enigmatic group of aquatic species called chytrids that form swimming spores, the basidiomycetes, ascomycetes, and zygomycetes constitute a kingdom of organisms separate from the animals and plants.[5]

Now we come to the single most perplexing concept in the study of fungi. Two or more apparently different fungi can be alternative embodiments of the same thing. Mushrooms, morels, truffles, and other kinds of large fruiting bodies are products of a sexual union between compatible strains, or sexes, of the same species. (I'm deliberately avoiding the many complexities introduced by sexual reproduction involving multiple mates and even hybrid mushrooms formed from a couple of different species. For now, I'm striving for an unambiguous scientific description of a mold.) Fruiting bodies are sexual organs that produce special kinds of spores called basidiospores and ascospores. When these microscopic seed-like structures germinate, they produce a mycelium of hyphae that penetrates a food source. If one mycelium meets another compatible mycelium, they may fuse, form a new fruiting body, and funnel their mingled genes into billions of spores. This is fungal sex. But fungal dispersal isn't dependent on this bacchanalia. Even the solitary mycelium is capable of moving on when it runs out of food. It does so by producing spores without any cooperation with a mate: Asexual spores develop at the tips of hyphae without the need for a fruiting body. When either phase is viewed alone—the one with fruiting bodies or the other that produces clouds of asexual spores—it appears to be a unique species. The link between the sexual and asexual beings is hidden at the genetic level. A single set of instructions is sufficient to create two types of spore-producing organism (and sometimes more than two). Most of the time when the term mold is used, we are referring to one of these asexual stages.

This is a tough concept, because we are unaccustomed to the idea of a single plant or animal existing in different states (though many of them do). Male and female humans are strains of the same species, but imagine, for the sake of this explanation, that we don't produce eggs or sperm cells. Instead, we birth "spores" from time to time that develop into distinctive animals that spend their lives hopping around in trees and doing little else but chugging-out eggs or sperm. Each time an egg is fertilized by a compatible sperm, a fruiting body shaped as an enormous mushroom emerges and dangles baby boys and girls from its gills. Forget the gooseberry bush.

I swear I have not been taking anything. I'm trying to illustrate that a mold is a stage within a larger fungal life cycle. Black molds, like most molds, are asexual phases of ascomycete fungi. This is obvious from their genes, and, for some of them, the ascomycete connection is demonstrated because they will produce ascospores from ascomycete fruiting bodies when a couple of strains (sexes) meet on a culture plate. *Stachybotrys chartarum* and many other species never do this, at least not in the laboratory. They keep growing and producing conidia and never mate. Mycologists think that these fungi have dispensed with sex. They are so successful at reproducing in solitude that different strains of a single species can diverge from one another so that they can no longer fuse, shuffle their genes around, and disperse the resulting progeny in sexual spores. In other words, new asexual species can evolve and quit the mating game entirely. The production of identical copies of oneself forever and ever runs counter to the ubiquity of sex in the natural world. Fungi mitigate the impact of this clonal way of life on genetic variation by employing asexual mechanisms for recombining genes in novel combinations. This is called parasexuality. The absence of sex among such widely distributed organisms should probably be viewed as a consequence of prosperity.

Stachybotrys spores are called conidia. They are single cells, shaped like footballs, and their surface becomes roughened with short, ropey strands as they mature. The chemical nature of the strands is not clear, but the toxins generated by some strains of the fungus are likely to reside in them. When the spores are viewed with a light microscope, they can appear spiny (see figure 1.1), but this is an optical illusion caused by those surface ripples. (Think about the silhouette of a curved sheet of corrugated iron.) The patterned exterior of the spore only becomes visible when highly magnified with a scanning electron microscope (plate 2). This instrument— developed 50 years ago—energizes the sample by playing a stream of electrons over its surface, which then discharges a spray of electrons from the spore. The microscope captures these emitted electrons and uses them to display a picture of the specimen's surface on a miniature TV screen.

Stachybotrys spores are produced by cells called phialides (*fy-a-lides*) that are bunched together at the tips of spore stalks or conidiophores—filaments that stick up rather than penetrating the food like the hyphae that form rest of the colony (plate 3). Each of the phialides is shaped like a bowling pin and extrudes a succession of spores like toothpaste from a tube. Initially, the spores are smooth; the appearance of the surface ripples is a stamp of maturity. With enormous numbers of phialides operating from a forest of conidiophores, masses of the black spores accumulate above a particular colony. The spores are surrounded by a paste or mucilage, and as this dries, they remain clumped on top of the colony. Along with the pigmentation of the hyphae, it is the density of black spores that makes an area of mold damage visible in a home. Because the spores are stuck to the surface of the colony, some kind of disturbance must occur before the conidia are dispersed into the air. Obvious mechanisms would be air movement and building vibration. These were tested by researchers at the University of Cincinnati and at the U.S. EPA laboratory in Research Triangle Park in North Carolina. The group in Cincinnati looked at common indoor molds that produce dry spores, and Marc Menetrez at the EPA studied the sticky spores of *Stachybotrys*.[6] The experiments involved a special chamber in which air could be blown over fungal colonies growing on pieces of ceiling tiles or other materials. Mechanical disturbance was controlled by battering the tile with a vibrator that oscillated up and down when activated by an electromagnet. The effects of different treatments were measured by counting spores with an instrument called an optical particle counter as they wafted from the tile surface. The researchers studied airspeeds ranging from typical air currents measured in rooms to the much faster flow rates inside ventilation ducts. It was more difficult to figure out what level of vibration would correspond to the kinds of disturbance produced by normal family activities. Some people slam doors from time to time; others settle arguments with a sledgehammer. Traffic rumbling past a house, aircraft streaking overhead, people falling out of bed, and children (no matter what they are doing) all cause colossal vibrations. Considering this universe of complexity, the researchers settled on a low frequency that would be expected to shake a ceiling tile. These beautiful experiments established that even low airspeeds were effective at driving spores from the ceiling tiles, and that vibration increased the concentration of spores flying past the particle counter. Fewer conidia of *Stachybotrys* were liberated into the air than those of the dry-spored molds, but the experiments did confirm that the beast isn't irrevocably glued to the tiles.

This result is not surprising, because spores of *Stachybotrys* are captured when air samples are taken from contaminated buildings. But laboratory experiments on spore dispersal are critical if we are to understand how and when mold spores move around our living spaces. Spores go unnoticed, of course, unless a room is lit through a single high window, or a narrow gap between curtains, whereupon the air's transparency is transformed into a cloud of frenzied specks glinting as they rise and fall in the spotlight. Among the motes of fungal origin float pollen grains, skin flecks, animal dander, carpet fibers, and mite feces (to which I'll return in the next chapter). If, like me, you are a hypochondriac, I suggest you don't spend much time pondering the filth in our air supply. Trust the housekeeping skills of your lungs and worry about other things. (This morning, I'm writing in a scrupulously clean college library, but thanks to my gymnastic imagination I might as well be typing in an African asbestos mine.)

I have explained that a single fungus can produce different growth forms depending on the stage of its life cycle. The environment can also have a profound effect upon the appearance of a fungus. Satiated or starving, despite obvious weight differences, all human beings have a similar form. By contrast, the shape of fungal cells and the way that they develop can show radical changes in relation to nutrient availability. The British mycologist C. T. Ingold discovered many new growth forms and styles of spore production in well-known fungi, simply by growing them on agar medium in which the nutrients were diluted far below the conventional level. (This was one of his retirement projects conducted in a home laboratory.) In addition to changes in appearance, food availability also affects the metabolic activity of fungi, so a single indoor mold can produce lots of toxins when it is soaked, and none when water becomes scarce (or vice versa). These considerations play an important role in identifying indoor molds and in judging their toxicity.

I need to introduce a couple of taxonomic terms here: genus and species. *Stachybotrys* is a fungal genus, just as *Rattus* is a genus of mammals. *Stachybotrys chartarum* is one of several species—some say 11—circumscribed by the genus; *Rattus norvegicus* is the Latin name for the sewer rat (which is even less welcome in most homes than black mold). Another fungal genus called *Memnoniella* (*mem-known-ee-ela*) comprises four species of mold. *Memnoniella* produces spores at the tips of phialides in the same way as *Stachybotrys*, and some species in both genera are adept at cellulose breakdown and develop as indoor molds. One difference between them is that the conidia of *Memnoniella* species are almost spherical (figure 2.1), contrasting with the football-shaped spores of *Stachybotrys*. But these differences are most obvious only when the molds are grown on rich culture medium containing corn meal. When *Memnoniella echinata* is grown on agar containing cellulose, it begins generating more elongated spores that are difficult to distinguish from those of *Stachybotrys*. To decide whether we are dealing with two distinct mold genera or a single entity, researchers at the EPA in Cincinnati compared DNA sequences from cells of both genera. They concluded that there was no scientific reason to maintain the name *Memnoniella*.[7] *Memnoniella echinata* is now called *Stachybotrys echinata*. I apologize for making your eyelids very heavy—I must admit I was beginning to drool myself—and I promise to stay away from taxonomy for the rest of the book. But this little disquisition conveyed some useful information. Molds are highly plastic in their growth forms (or morphology) and can be confused with one an-

Figure 2.1
The black mold *Memnoniella echinata*. Some mycologists believe that this fungus should be classified as a species of *Stachybotrys*. (From M. B. Ellis, *Dematiaceous Hyphomycetes*, Wallingford, United Kingdom: Commonwealth Mycological Institute, 1971. Reprinted with permission.)

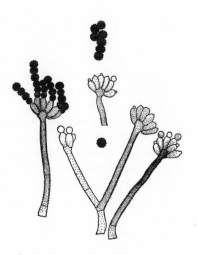

other very easily when examined with a microscope. Even the experts disagree; despite the EPA study, some refuse to abandon the name *Memnoniella*.

Mold identification is becoming more of a science than an art thanks to the introduction of genetic methods, and this has important implications for surveying molds in homes. Richard Haugland and Steve Vesper were the lead authors of the *Memnoniella* study and have since licensed a technique that discriminates between 100 different species of potentially dangerous fungi. Haugland is a quiet man with a wealth of experience working with fungi; his colleague Vesper is a live wire who bursts with anecdotes about *Stachybotrys* and its toxins. They are world leaders in this field, though working for the EPA they have been relegated to a small laboratory space by researchers studying anthrax. In any case, their detection method uses the polymerase chain reaction (PCR) that many people first heard about during the trial of O. J. Simpson in the 1990s.[8] The PCR method allows the investigator (of fungi, or unrepentant murderers) to amplify tiny samples of genetic material into quantities that can be conveniently sequenced. Molecular identification of molds relies upon small differences between certain sequences in their genetic makeup and the method is sensitive enough to determine whether *Stachybotrys* is present in a room even if it is not actively growing or its spores are dead. As I have mentioned earlier, the mere identification of *Stachybotrys* in a home does not mean that the residents are in danger.

Stachybotrys chartarum was first isolated in 1837 from wallpaper in Prague. It seems to enjoy a global distribution in buildings; and in addition to its prevalence on wet paper products, mycologists have isolated the mold from moldy leaves, soil, desert sand, the gut contents of bees, seeds, cat hair, and feathers. In the last 2 years, the fungus showed up in a couple of new locations: on soybean roots, where it may have been acting as a weak pathogen,[9] and from sea turtle cloacae, from which it was extracted by a team of Australian animal gynecologists.[10] Surveys of indoor molds have often listed *Stachybotrys* as an infrequent inhabitant relative to other molds, with some researchers reporting it in as few as 3 percent of samples taken from damp buildings.[11] More recently, however, mycologists have recognized that the standard culture methods are at fault in underrepresenting *Stachybotrys*, which further emphasizes the importance of the new PCR-based method of identification. Related investigations have also suggested that *Stachybotrys chartarum* may encompass two genetically distinct species that produce different kinds of toxins.[12] If this finding is borne out by further research, mycologists will have to coin a name for a new species.

The prestige of *Stachybotrys* is a purely human illusion: Other molds are far more common and are immeasurably more important to life on Earth as agents of decay. Mycologists have described 74,000 fungal species,[13] a number that means nothing without some benchmarks. I'll look outside my shed for a moment for inspiration. I rescued a toad from one of our cats last night: There are 4,700 or so species of frogs and toads, and 35 kinds of cats besides the domestic variety. A prolonged drought is stressing the trees in my yard, and even though it's early September, the leaves are falling: Botanists have described a quarter of a million types of green plant. The cooler weather hasn't ended the songs of the cicadas yet; insects are ridiculously diverse: More than 750,000 species are known. So for every fungus identified by a mycologist, representatives of 10 insect species are pinned in museum cases. The fungi that we know, however, are among the species most easily grown in culture dishes and live in the places where mycologists have worked for the last couple of centuries. If we consider the concentration of biodiversity in the tropics, and the fact that every insect, plant, and other kind of organism interacts with multiple kinds of fungi—pathogens and saprobes—then estimates of the actual number of fungi quickly exceed a million. Perhaps 1.5 million are waiting for a name tag.[14]

In the 1970s, mycologist Martin Ellis catalogued hundreds of black mold species in a monumental work titled *Dematiaceous Hyphomycetes*, and hundreds more were laid bare in his sober-titled sequel, *More Dematiaceous Hyphomycetes*.[15] Most people would find these volumes horrifically tedious, but a quick flip through more than a thousand pages of drawings and formal descriptions provides an immediate and lasting impression of the incredible diversity of black molds. I'll crack open the books at random and describe the molds (figure 2.2). *Periconia curta* grows on sedges in Europe, producing half-millimeter-diameter heads of conidia on single stalks that are visible without a microscope. *Blastodictys hibisci* resides on hibiscus plants in Uganda where it was discovered in 1943. The spores of this species are large, multicellular blobs, each furnished with a single, upwardly projecting horn. It is a microscopic unicorn. Another, *Virgaria nigra,* has a global distribution, was first described in 1816, grows on the bark of numerous tree species, and produces its spores on branches that look like pimpled "for extra pleasure"

Figure 2.2
A selection of black molds. (a) *Periconia curta*. (b) *Blastodictys hibisci*.
(c) *Virgaria nigra*. (From M. B. Ellis, *Dematiaceous Hyphomycetes*,
Wallingford, United Kingdom: Commonwealth Mycological Institute,
1971, and M. B. Ellis, *More Dematiaceous Hyphomycetes*, Wallingford,
United Kingdom: Commonwealth Mycological Institute, 1976.
Reprinted with permission.)

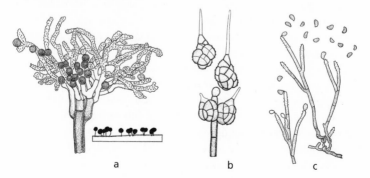

a   b   c

dildos. In the years since the publication of the Ellis books, new black molds have been added to the catalog as swiftly as mycologists can write their descriptions. The total number is unclear, but must run to tens of thousands.

My house in Ohio is a conventional suburban American home built in 1994, with three bedrooms, two and a half bathrooms, an indoor ski slope (not really), and a library addition. It does not have a basement, which is a major disincentive for molds that get very excited by the availability of a dank den. The property has suffered a few episodes of water damage. Two floods have been caused by my teenagers plugging an upstairs toilet, whose overflow seeped through the kitchen ceiling. On both occasions the wet flooring was dried and the ceiling drywall was replaced. Other inundations were more serious because they came from outside. Due to the property's location at the low end of the neighborhood, one part of the house was flooded before we decided to pay a vast sum to a contractor to sink a pipe large enough to drain a lake. Despite the water damage, and spore-dispersing vibrations caused by my screams, we have never had a mold problem. This is a full telling of the history of the house. It is perfectly dry, and I would most certainly buy the place without hesitation if I hadn't done so already. With this in mind, I decided to conduct a very detailed mold survey of my mansionette.

Traces of black molds were obvious on shower curtains, on the drywall above one of the showers, around sink drains, and, most lavishly, speckling the polystyrene insulation covering a floor-standing water cooler. I had never noticed a black crust around one of the shower faucets before I embarked on my probe of indoor fungi, but now this growth assumed the proportions of a major threat to homeland security. After taking a sample of the crust with a cotton swab, I expunged the rest with brush and bleach (without the protection of a biohazard suit). In the lab, the grunge specimens were examined under the microscope, and the tips of the swabs were dabbed onto agar medium to allow further growth in an incubator. The crust around the faucet was fashioned from a thick raft of entangled hyphae of the mold *Cladosporium*, whose cells were filled with translucent oil droplets. Presumably, this microbe made its living by feeding on the lipid-enriched residue of soap and shampoo. Interestingly, no bacteria were visible. The fungal crust may have kept them at bay by secreting antibiotics.

Another fungus that I encountered produced large, lumpy conidia with multiple compartments. On culture plates, its mycelium turned to a tar when it became covered with spores. Consulting Ellis and other sources, I determined that this one was a *Monodictys* that is often found on rotten wood, linoleum, paper, and sacking. It is easy to be mistaken about the cause of a discolored area in a house, because spores of other fungi floating in the air may contaminate the sample and may be better at growing on agar than the mold that blackened the wall. In other words, my *Mondictys* culture might have originated from spores that drifted into the house from a pile of logs in the yard. To dismiss this possibility, it is important to examine samples promptly before culturing. In the case of the *Monodictys*, its hyphae and spores were clumped on the tip of the swab, so it was obvious that the fungus had been growing furiously in a greasy ring around one of the sink drains. A new edition of Ellis—*Even More Dematiaceous Hyphomycetes*, perhaps—should add that *Monodictys* also likes Ohio bathrooms. It is difficult to say whether other molds that grew up on the same culture plates had been active in defiling my bathroom fixtures.

Now to the lip balm mentioned in the preface. The cosmetic product was heavily contaminated with the black mold *Aspergillus niger* (plate 4). One or more spores of this fungus may have alighted on the mixture during manufacture or packaging, or could have settled after the little container was opened. Unlike the *Cladosporium* and *Monodictys* cultured from my home, the *Aspergillus* spores were associated with a disagreeable effect—severe lip irritation—and the credibility of a causal relationship was strengthened by the observation that the symptoms were relieved by discontinuing use of the moldy salve. Join me in a thought experiment. If something worse than lip irritation had occurred, would there have been any merit to a lawsuit against the manufacturer of the lip balm? Such a lawsuit would be inherently weak, because we could not show any proof that the fungus caused Diana's injury. To validate the noxiousness of the product, it would be necessary to compare the effects of contaminated and uncontaminated samples on a number of people. Without this information, the manufacturer could claim that the plaintiff may have had a persistent problem with lip irritation that had nothing to do with their product. A jury might conclude that there was reasonable doubt that the mold was the cause, or, even if it was the irritant, that the manufacturer wasn't responsible for natural contamination of its product after the container was opened. Likewise, without a controlled study, it could have been the chemicals in the lip balm itself that provoked the burning sensation, which would necessitate a different tack on the part of the prosecution.

A burning sensation is a trivial consequence of a black mold encounter, particularly since *Aspergillus niger* can be a killer. Mycologists have described well over 100 species of *Aspergillus*, and several of these cause human infections. *Aspergillus niger* is the most widespread species of *Aspergillus* and is one of three species responsible for a disease called invasive pulmonary aspergillosis. The fungus is a natural inhabitant of soils, and its spores can be isolated from air samples taken outdoors and indoors, even from the air circulating around a hospital. Like *Stachybotrys*, it generates masses of spores from the specialized cells called phialides. The phialides of *Aspergillus* are arrayed on stalks (conidiophores) with bulbous heads. It is a beautiful microorganism when confined to a Petri dish. Each phialide produces a long chain of conidia, and these take on the appearance of dreadlocks under the microscope. The spores are not produced in a slime like those of *Stachybotrys*, so they disperse in the air with very little disturbance.

Aspergillus conidia are probably the most commonly inhaled spores, but aspergillosis is very rare in people with healthy immune systems. This caveat, always offered in textbooks of medical mycology, is cold comfort for anyone with an impaired immune response. When it does occur, invasive aspergillosis is usually lethal. Once the lungs are breached, the mold can spread to other organs and can even be carried through the bloodstream. If the fungus penetrates the walls of blood vessels supplying the brain, the ensuing damage can cause a stroke. Other causes of death include hemorrhaging of the lungs, stomach, or intestine. *Aspergillus niger* is under additional suspicion because some strains of the fungus—perhaps as many as 10 percent—are known to produce ochratoxin A and other potentially harmful metabolites.[16] The health impact of these compounds, if any, has so far escaped the detailed scrutiny enjoyed by the mycotoxins carried by spores of *Stachybotrys*.

Recall that I have been describing black molds found in my home. This is a clean house without any persistent water damage—barely a damp spot. It doesn't get any better than this. Acceptance of some mold growth in a home is a very beneficial attitude; quantity is a decisive issue, and my house passes this test with ease. In addition to the sense of well-being that comes with a more relaxed attitude toward the unavoidable microbes that share our living space, there is growing evidence that a disinfected home is an unhealthy environment for a child. Without early challenges to our immune systems, we become more likely to suffer from asthma and other kinds of autoimmune dysfunction. I'll discuss the immune response to fungi in the next chapter.

Before I move beyond my property, I want to examine my writing shed again, because without any encouragement on my part, this has become a living temple to mycology. For a long time, I assumed that the discoloration was produced by dust blown and splashed from the surrounding flower beds, because the densest stain covered the bottom half of the planking. (You would think that I would have looked more closely, but then I am the Professor of Botany known for shutting his own head in a car door and, most recently, for allowing a flask filled with boiling horse feces to explode in the laboratory.) Eventual closer examination revealed that the blackness was caused by millions of mold colonies, many assuming a lens-shaped outline because they extruded from fissures called rays that are a part of the anatomy of the wood. The reason for the location of most of the growth was that the bottom part of the shed is not protected by the roof overhang, and is the dampest part of the building. What irony to be writing about specific types of fungi and to be surrounded by their colonies! A number of fungi are responsible for the discoloration. Some penetrate the outermost layer of wood, others weave around in the coat of water-seal and effuse spores wherever they emerge at the surface. Most rampant is a species of *Epicoccum*, a fungus that invades lesions on plants formed by other fungi; Ellis calls it a secondary invader. Because my shed is made from dead wood, the relocation from plants in the yard was a no-brainer. With the help of my friend Chris Wood (expert on bioinformatics and Ohio's reigning King of Power Tools), I had built a refuge for myself and unintentionally provided dinner for billions of *Epicoccum* hyphae.

The mammalian inhabitant of the shed represents another potential food source for molds. Plenty of them would relish more protein in their diet. In addition to *Aspergillus niger*, 60 or more other species of black mold that cause human disease have been reported in the scientific literature. Infections range in seriousness from fungi that form nodules on hair shafts and discolor small patches of skin to those that cause incurable brain infections. Black mold infections of the brain are rare. It is thought that invasion begins in the lung and that the fungi spread to the central nervous system through the bloodstream. The fungi form abscesses in almost any location within the brain, and the infection is "silent" until the development of neurological symptoms that include headache, paralysis, confusion, impaired vision, slurred speech, seizures, and convulsions. *Wangiella* is the most common of the neurotropic (brain-loving) black molds in the Far East; *Ramichloridium mackenziei* is limited to the Middle East; and a beast called *Cladophialophora bantiana* stalks westerners. All three fungi are closely related (they belong to the same taxonomic family) and are remarkable in their ability to cause lethal diseases in otherwise healthy patients. A fourth black mold, *Ochroconis gallopava*, also causes brain infections, but this one is forceful only when it encounters someone with a damaged immune system. This is a strange fungus among strange fungi. *Ochroconis* can tolerate temperatures as high as 50 degrees Celsius, which classifies it as a thermophile and allows the fungus to inhabit the fringes of the hot springs of Yellowstone National Park. Though it hasn't infected any tourists, its partiality to brains has led it to attack wild turkeys in the park.[17] My friend Joan Henson at Montana State University sent me a culture of this mold some years ago. I haven't plucked up enough courage to open the culture tube yet, so it sits on a shelf, looking like an oil slick on a little slant of agar.

Black molds, including *Aspergillus niger* and species of *Alternaria*, *Curvularia*, and my shed-infesting *Epicoccum*, are among the more than 50 fungi now implicated in a form of chronic sinusitis known as allergic fungal sinusitis or AFS. Chronic sinusitis is a very common problem, but fungi were thought to be involved in only 10 percent or so of cases until a Mayo Clinic study found that more than 90 percent of patients harbored multiple species of mold in their nasal passages![18] The hallmark of AFS is allergic mucin, a black material with the consistency of peanut butter, which is removed during surgery to clear nasal blockages. Besides surgical intervention, treatment of chronic sinusitis has relied on the use of steroids to reduce inflammation. Now that fungi are recognized as a feature of most sinusitis cases, the use of antifungal medications is under investigation.[19] *Alternaria* was found in the nasal passage mucus of 40 percent of the patients diagnosed with AFS at the Mayo. Its spores are very large, 10 times longer than those of *Stachybotrys*, and so they might be expected to become trapped in the nose rather than passing down into the lungs. In response to contact with these spores, the airways of sensitive individuals would constrict, increasing the efficiency of the nose as a spore trap. This allergic response would also prevent clearance by sneezing or nose blowing, prolonging exposure to the spores that caused the airway constriction in the first place. This explanation is very logical, but the Mayo Clinic study showed that AFS lacks the cellular hallmarks of a classic allergy. Based on their observations, the researchers suggest renaming AFS as eosinophilic fungal rhinosinusitis. Occasionally, fungi in the nasal sinuses form dense spheres of hyphae—termed "fungus balls" by clinicians—and, in very rare cases, the mold can spread through the bloodstream to the brain. The latter outcome is fatal.

Black molds on wood, and black molds in brains and noses. Where else? Everywhere, on everything, living or dead. Plants and photosynthetic bacteria coerce the carbon atoms from atmospheric carbon dioxide into glucose molecules and then manufacture all the other carbohydrates, plus fats, proteins, and nucleic acids they need. Certain types of bacteria have alternative tricks for making their biological molecules, but one way or another everything must create or steal the things it needs. Black molds grow in surprising abundance on natural rock surfaces and on cut rock when it is turned into buildings, tombstones, and other monuments to human brilliance. This seems incredible because marble, granite, and sandstone contain little or no organic matter. A controversial resolution to this paradox is mentioned from time to time at scientific meetings but is dismissed for lack of evidence (and because the suggestion is treated as heresy). This is the idea that fungi can absorb carbon dioxide from the air and use its carbon atoms to build biomolecules through a non-photosynthetic mechanism. Some fungi will grow under conditions of extreme nutrient limitation, and they do appear to incorporate some carbon from the surrounding air.[20] These are intriguing experiments, but they do not establish that this cryptic biochemical process is capable of feeding the fungus. For rock-dwelling black molds, a more plausible suggestion—or at least a more widely accepted one—is that they feed upon hydrocarbon-rich pollutants from automobile emissions, dust, and animal excretions.[21] So there you have it: impress your friends—that soot on the office building is alive.

Even more astonishing is the discovery that the molds prevailing in these apparently inhospitable places are close relatives of the species that infect humans.[22] You wouldn't have guessed that we tasted anything like grimy sandstone would you? We don't, but there is a way to make sense of the incredible range of food sources exploited by black molds. The evolution of melanin synthesis among the ancestors of today's black molds enhanced their fitness so greatly, and in so many ways, that they discovered all kinds of habitats in which competition for space and nutrients from other organisms, including non-melanized fungi, was absent. Nothing else could bloom on leaves or logs exposed to the sun and wind; without the protection offered by melanin, all the transparent fungi shriveled for lack of water and burnt in the ultraviolet light. The black molds also joined a select fraternity of parasites capable of beating the immune defenses of unhealthy animals, and sometimes prevailed over healthy animals, too. There wasn't much competition for the blisteringly hot surface of exposed rocks either. All over the planet, shrouded by their melanism, black molds escaped from the environmental strictures imposed on other organisms. The fact that they feed on such a diverse selection of food substances is not surprising to mycologists. All fungi possess a considerable catalog of enzymes and can tailor their secretions to make use of whatever food they encounter in their environment. Apparently, specialized human pathogens will grow happily on plant tissues if this is what they are given. Thus, blackness is my explanation for the panoramic tastes of these molds, for their enduring success, and for their menace to humanity.

I was born with inflammation of the lungs,

and of everything else, I believe,

that was capable of inflammation.

—Dickens, *Hard Times* (1854)

Carpet Monsters

Mold spores provoke allergies and asthma, and those that carry mycotoxins can do much worse. My early childhood was defined, in part, by asthma. The illness kept me at home on many school days, which pigeonholed me as a delicate boy, an undesirable message that was stamped into my subconscious and throws an enduring shadow. (I have already mentioned my dreadful hypochondria.) Invisible monsters in carpet dust came close to asphyxiating me, and I spent some time in an oxygen tent in a hospital. Thankfully, the severest symptoms were shortlived. The introduction of cromolyn sodium (Intal) inhalers in the early 1970s—"spinhalers" from which one sucked the powdered drug, before it was formulated for delivery by a propellant-driven inhaler—alleviated my disability and gave me a lifelong love of pharmaceutical companies. Thank you, Fisons Company, U.K.,[1] Sincerely: You gave me a life. In common with many expatriates, moving to America has allowed me to escape my European irritants (particulate and human). England remains a very allergenic place for me, and I cannot survive there for more than a week without a daily puff of bronchodilator. It is ironic to be beginning this chapter today, because I have soaked a pile of handkerchiefs by sneezing since I awoke this morning. The black mold surrounding me on the shed looks the same as it has for months, so I suspect that the culprits are masses of other kinds of fungal spores that became airborne yesterday as a cold front moved through the Ohio Valley. Indeed, half the city seems to be sneezing. Sales of Allegra must be astronomical.

There are good reasons to think that fungal spores are the cause of my allergies and for the struggles of millions of fellow sufferers. A 2002 study in the *British Medical Journal* concluded that a patient's sensitivity to airborne molds was a far better predictor of the severity of their asthma than their response to pollen or cat hair.[2] Part of the explanation is that most fungal spores are smaller than pollen grains, making them more likely to elude the sticky walls of the upper airways, and reach the deeper parts of the lungs. So here's another irony: My childhood was profoundly affected by fungal spores, but without being aware of this I became a professional mycologist. I even failed to think of this when I embarked on a book about molds. Perhaps my slowness can be blamed on all the school I missed while gasping for oxygen, but given the brilliant careers of other asthmatics,[3] this isn't an ironclad exegesis.

An important feature of the allergic response is that spores and other particles are carriers of minute quantities of particular proteins called allergens, which cause allergies. Picture a football dusted with powdered sugar: The sugar represents the allergens, the football is the rest of the spore which is invisible to the immune system. When spores enter the airways of someone with allergies—referred to as an atopic—soluble proteins on the spore surface dissolve in the mucus that coats the internal labyrinth. The same thing happens when pollen grains, the feces of cockroaches and dust mites (a common scourge of asthmatics), and other allergenic particles are inhaled. Scientists think that atopics misidentify these foreign proteins as the signatures of potential parasites and infectious agents and mount a defensive response that underlies the symptoms of asthma.

Four cell types play the lead roles in mediating the body's response to foreign proteins encountered in the lungs and intestine, and at the skin surface: dendritic cells, T cells, B cells, and mast cells (figure 3.1). (Any immunologists should skip the next few pages if they want to avoid gnashing their teeth at my efforts to elucidate this intricate mechanism.) Dendritic cells are shaped like starfish.

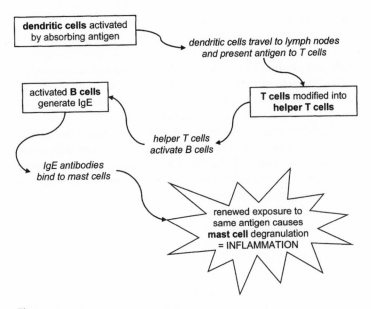

Figure 3.1
Diagram illustrating interactions between immune cells
in an allergic response.

They live in most of our tissues and function as scouts, looking for molecules that signify the presence of disease-causing organisms and viruses that might be invading the body. When a dendritic cell is nudged by a bacterium, it responds by engulfing the bug and degrading it within its cytoplasm. Proteins derived from the fragmented bacterial cells are then carried to the surface of the dendritic cell. In addition to processing whole bacteria, the dendritic cells also deal with viral particles and free molecules, including allergenic proteins originating from dust mites and spores. Dendritic cells are activated by feeding, whereupon they travel to the nearest lymph node. Arriving at the lymph node, they behave as "professional antigen-presenting cells" by displaying the processed proteins—antigens—to cells called T lymphocytes, or T cells. In response to the gift, the T cells become modified into helper T cells, which are capable of inducing another type of immune cell called the B lymphocyte to produce antibodies. In the allergic response, B cells generate a particular type of antibody called immunoglobulin E, or IgE.[4] The IgE molecules are sculpted to bind the original molecules presented by the dendritic cells. If the original challenge was mite feces, B cells will produce mite feces-specific IgE. Through this cascade of cellular interactions between dendritic cells, T cells, and B cells, the body primes itself to recognize the foreign material. One more cell type and we'll get the patient wheezing, I promise.

The final player is the mast cell. Mast cells reside in epithelial tissues close to blood vessels. Here's what they do. Mast cells absorb those IgE molecules and use them to decorate their cell surfaces. Each cell may bear an astonishing quarter of a million IgE molecules on its surface, making it responsive to a wide spectrum of antigens. When an appropriate antigen is encountered, the irritating protein binds to the mast cells by forming cross links between adjacent IgE molecules. This enrages the mast cells—even textbooks of immunology tend to anthropomorphize this aspect of allergy—and they degranulate. Degranulation describes the evacuation of granules that lie beneath the cell membrane. The granules contain inflammatory agents, including the infamous histamine. Histamine release causes an increase in blood flow and blood vessel permeability in the affected regions. This is called a local inflammatory response. Mast cells also produce other compounds that cause smooth muscle contraction and stimulate mucus production.[5] These reactions serve a crucial function by alerting the immune system that an infection may be in progress. Other mechanisms allow the immune system to fight bacteria and infectious agents, but the mast cells are an irreplaceable component of our defenses.

Mast cells can be activated in atopics whether the allergen is inhaled, brushed against, eaten, or injected. Allergic rhinitis or hay fever, allergic conjunctivitis, skin allergies, food allergies, and anaphylactic shock all share similar immunological mechanisms with asthma. Hay fever and allergic conjunctivitis are caused by airborne allergens, particularly pollen and mold spores. In the allergic response to an insect bite, local injection of an allergen causes direct stimulation of mast cells in the skin. The resulting redness is referred to as a wheel-and-flare reaction. Urticaria or hives refers to more widespread skin irritation caused when an allergen is ingested and reaches the skin through the bloodstream. Allergy to penicillin, for example, is often manifested as urticaria. Allergy to peanuts or shellfish can also result in a life-threatening anaphylactic shock involving severe asthma and cardiovascular collapse. Like all of these conditions, the persistent childhood skin rash known as eczema or atopic dermatitis involves T cells and IgE, but is otherwise poorly understood.

The most potent allergens appear to be enzymes, particularly those that catalyze the destruction of other proteins. Protein-degrading enzymes are called proteases or proteinases. The identification of an allergen from dust mites called Der p 1 offers some tantalizing insights about the immune system. Dressed in mucus, a protective layer of epithelial cells forms the convoluted surface of the nasal passages and lungs. The epithelial cells are tethered to one another—side by side—via protein links so that they create a continuous barrier. The mite protein Der p 1 cleaves one of the components of the junction between the cells, disrupting the barrier and allowing itself and other proteins in the mite feces to breach the epithelium. Beneath the epithelium, these proteins interact with the dendritic cells and initiate the immunological cascade I have just described. Many allergens are proteases, but there are exceptions. Certain proteins that function as inhibitors of enzymes are also good at provoking an immune response. But the fact that proteins like Der p 1 are allergenic points to an interesting explanation for the immunological mechanism that is corrupted in atopics. Rather than protecting the body against disease-causing bacteria, fungi, or viruses, there is evidence that the IgE-based immunological reaction evolved as a protection against parasitic worms.[6] The common theme that likens protease allergens and parasitic worms is their mutual damage to the protective epithelium that lines the intestine and lungs, and also forms the skin barrier. In addition, parasites produce their own proteases, which may further diminish the distinction between inanimate mite feces and a live worm. Because the dendritic cells lie just beneath the epithelium, they are perfectly situated to recognize the earliest stages of colonization. So when a protease from a dust mite penetrates the lung lining, the IgE response is mobilized to combat a nonexistent attack by parasites. According to this explanation, allergies—or at least some forms of allergy—can be viewed as a perversion of a response designed to deal with parasite infestations. In an allergic response there is nothing living that can be disarmed but oneself.

A dismissive attitude toward intestinal worms is understandable for anyone capable of paying $2 for a cup of coffee, but worms and other parasites still reside in the guts of billions of our species.[7] Only very recently, in biological terms, have sanitation and modern medicine succeeded in abolishing a panoply of parasites including roundworms, hookworms, flukes, tapeworms, and protists (like the malarial parasite). Even though it totes a kilogram of diverse bacteria, my intestine is a place of unprecedented purity compared with the teeming bowels of my ancestors. I live with the confidence of knowing that nothing that lives inside me is visible without a microscope. Fortunately though, my immune system will not forget how to deal with worms any time soon, which will be useful if I ever end up groveling in a cave and drinking water from a muddy puddle just like my great-great-great-etc.-grandparents. The absence of worms has also been blamed for inflammatory illnesses such as Crohn's disease and ulcerative colitis. In a seemingly perverse approach to treating these ailments, Dr. Joel Weinstock at the University of Iowa has pioneered the deliberate ingestion of parasites. According to Joel, many of his patients have reported dramatic improvement in their conditions after swallowing eggs of the pig whipworm, *Trichuris suis*. The whipworms cause no disease symptoms and are expelled from the body a few weeks after the patients' autoimmune responses have been canceled.

There are many different drug therapies for treating asthma and other so-called type I hypersensitive reactions. Cromolyn sodium coats mast cells and prevents degranulation. It is an excellent preventive medicine but does nothing to stop an asthma attack once it is in progress. Bronchodilators are used for this purpose and function by relaxing contracted muscles to open constricted airways. Rather than preventing degranulation, antihistamines block histamine receptors and can be very effective in curbing skin inflammation. Pseudoephedrine is a decongestant that acts as a potent vasoconstrictor, reducing localized blood flow to decrease swelling in the airways. In cases of chronic allergies, corticosteroids—synthetic versions of steroid hormones—are used to modulate the expression of genes that control the inflammatory response. These are very powerful medications, but come with numerous side effects.

Alternative therapies are designed to reformat the immune system by modifying the T cell response to allergens. When the undifferentiated or naive T lymphocyte is presented with antigens, as I explained earlier, it becomes transformed into a helper T cell. What I saved for a while was the news that two types of helper cell can be generated. The first of these, designated T_H1, is responsible for combating infections by activating the macrophages that engulf and digest viral particles and the cells of bacteria and fungi. B cells interacting with these helper cells produce a different type of antibody called immunoglobulin G (IgG) or gamma globulin. This is the most common antibody type in the bloodstream and lymph, and it labels pathogens for destruction. The second kind of helper cell, T_H2, is the one I've described already which induces IgE production and is responsible for allergies. For reasons that are unclear, the concentration of IgE antibodies can be 10,000 times higher in the bloodstream of atopics versus the rest of the population. A treatment called desensitization is designed to shift the response to an allergen away from T_H2, IgE, and the resulting mast cell degranulation, toward the differentiation of T_H1 cells. This is done by injecting an atopic patient with increasing doses of a particular allergen; though the treatment can be effective, the possibility of a severe allergic reaction to the injected allergen is a serious concern. Sophisticated drugs that will directly skew the immune system in favor of T_H1 differentiation represent a holy grail for drug companies and are the subject of intensive research.

About 40 percent of the human population is atopic, and more than 50 million Americans suffer from allergic diseases. For reasons that we do not understand, the number of asthmatics is increasing at an alarming rate. A report published in 1995 estimated that 15 million Americans are asthmatic, which represents an astonishing 75 percent increase in prevalence since 1980![8] Asthma is the number one cause of hospital admissions for children, and more than 5,000 Americans die of the ailment every year. The disease has some geographical quirks that have been widely publicized. Children in cities with low levels of pollution are more likely to be asthmatic than children in polluted cities. In the United States, Maine has the highest percentage of asthmatics, Louisiana the lowest. Farther afield, asthma rates are exceedingly high in Australia and New Zealand and much lower in more heavily industrialized countries where the air is fouler. A study in Germany supported this distribution pattern. A comparison between Halle (a city with severe pollution) and Munich (with much cleaner air) found higher rates of asthma in Munich but more frequent nonallergic respiratory illnesses in Halle.[9] Clearly standard measurements of air quality are too crude to predict the hazard for an individual asthmatic.

The preponderance of allergies and autoimmune diseases in more affluent countries has led to the idea that the problem lies with our cleanliness and success in treating infectious diseases. This is called the hygiene hypothesis.[10] The most sensible idea discussed by allergists is that the best way to prepare the immune system is to expose it in its infancy to a broad spectrum of proteins. This teaches our defenses to attack the nasty things and ignore the daily fluff. It seems that the infant immune system can be formatted correctly—by a tapeworm, for example—or incorrectly by household dust. If mistakes are made during the formatting process, the immune system will make errors and become overly sensitive to innocuous particles. Children born on farms do not develop asthma as frequently as children born in cities, suggesting that contact with animals and casual exposure to their feces may be beneficial.[11] Household pets have already been given a stamp of approval by researchers who found that children in homes with cats or dogs obtained some protection from allergies.

Contrary to the hygiene hypothesis, some immunologists believe that early exposure to mold spores can result in a lifetime sensitivity to the same allergens and that the process may begin in the womb. The idea is that low levels of allergens filter across the placenta from mother to fetus and inform the virgin immune system in a way that prepares the individual for a lifetime of allergies. After birth, the baby probably inhales higher levels of the same allergens that it encountered in the womb. In susceptible individuals, the immune system becomes programmed to produce an allergic response to these materials, while in normal children the nonallergic response involving T_H1 begins to dominate the allergic T_H2 pathway, and the child is asthma free. There is evidence that viral infections and repeated exposure to fungal spores may bias the developing child toward lifetime atopy. This model helps explain my sensitivity to allergens in my birthplace, and the remission produced by emigration. Throwing all of the ingredients together, we can offer a recipe for asthma: Mix one cup genetic predisposition with a pinch of allergens in the womb, and bake until birth; after removing from the womb—and I already have a baby here which I prepared before the show—sprinkle the same allergens over the little angel,[12] but, ladies and gentlemen, please ensure that no parasitic worms get into the bowel; finally, cook at about 98 degrees Fahrenheit for 5 years or until breathing difficulties are apparent. (Then take your creation to the emergency room.)

Though environmental factors are critical determinants of a child's allergies, alterations to a human gene called *ADAM33* are associated with an increased risk of disease.[13] The gene specifies another of those proteases, and mutations in *ADAM33* are presumed to cause abnormalities in enzyme function. If the authors of the study are correct in supposing that the errant version of the enzyme is capable of damaging the airways in some fashion, then a further layer of genetic complexity must be added to the already perplexing disease mechanism. This is an example of "permanent airway remodeling," which has become an important concept in asthma research and therapy. The lungs of asthmatic patients whose inflammatory reactions are not properly controlled with medications eventually show signs of irreversible mechanical damage caused by thickening of membranes, tissue destruction, and loss of elasticity. Severe asthma is capable of transforming flexible bellows into useless leathery bags. For anyone with asthma, this is very bad news; for the legal profession, this may be a tremendous opportunity. Nobody would disagree that a moldy environment can induce an asthma attack in an allergic individual. So what? Take your medication. But think about it: Permanent airway remodeling means that chronic mold-induced asthma may cause unalterable debilitation and eventual lung failure. This elevates the illness to a very serious condition (which is something that physicians have been saying for years). If I were a landlord, I'd make sure that molds were evicted from my apartments. (Further confluence between mycology, medicine, and the law is addressed in chapter 6.)

That's enough of Immunology 101 to help you appreciate the complexity of the relationship between molds and allergies. A few decades ago, much of what I have told you about the interactions between immune cells in the allergic response was unknown. I remember listening to a few very vague lectures on the topic as a student in the early 1980s. Today, the study of allergy is at the forefront of biology and medicine, and articles published in the *Annual Review of Immunology* (a journal that deals with the whole field of immunology) are the most frequently cited studies in the scientific literature. But physicians have been remarking on the problems caused by mold inhalation for more than 200 years. I began the previous chapter with an eighteenth-century quote from James Bolton on the ubiquity of spores and their irritating effects. Bolton was thinking about the spore clouds emanating from mushrooms when he made this early connection between fungi and "Quinsies [inflammation of the throat] and coughs." A century later, Manchester physician Charles Blackley published his personal researches on "hay-fever or hay-asthma."[14] In his introduction to the subject, Blackley commented that "Hay-fever is said to be an aristocratic disease, and there can be no doubt that, if it is not wholly confined to the upper classes of society, it is rarely, if ever, met with but among the educated." This seemingly bizarre nineteenth-century viewpoint now makes some sense in light of the hygiene hypothesis, though if this is correct, it's obvious that plenty of poor, and poorly educated, children in cities also are being damaged by their lack of exposure to the dirt found on farms. Blackley suffered from allergies and became an unwitting experimental subject when he inhaled spores while examining colonies of a green-spored *Penicillium* and a black-spored *Chaetomium* that had grown up on wheat straw:

> the *odour* of the *Chaetomium* brought on nausea, faintness
> and giddiness on two separate occasions. By inhaling the
> spores of the *Penicillium*, in the involuntary experiment of
> which I have spoken, a severe attack of hoarseness, going
> on to complete aphonia [loss of voice], was brought on.
> This lasted for a couple of days, and ended in a sharp at-
> tack of the bronchial catarrh, which almost unfitted me for
> duty for a day or two.

He said that "The sensations caused by these two agents were so unpleasant that I have never cared to reproduce them."[15] (Blackley's labors in the cause of mycology are utterly eclipsed by Robert Remak, who deliberately infected himself with spores in 1842 to prove that ringworm was an infectious fungal disease!)

The human body can be likened to a spore sampler used by an industrial hygienist. After all, both suck spores from the air and deposit them on sticky surfaces (microscope slides or wet air passages). Humans are more efficient than electronic samplers, because they are highly mobile and sample the air every few seconds for as long as a century. This is a foolish calculation, but I can't stop myself from punching the numbers into a calculator. Let's assume that I'm going to live to age 80. If, on average, I have inhaled, and will continue to inhale, 10 spores per minute (or 14,400 spores per day), I will have sucked down 420 million spores by 2042. This may sound like a lot, and you may be expecting me to say something about the mountain of soot that might cast a shadow over my corpse. If so, you are forgetting just how small a thing is a spore. Assuming an average spore mass of less than a billionth of 1 gram, we arrive at a lifetime spore sample that tips the scales at 0.04 grams, equivalent to the weight of two grains of rice.[16] An agricultural worker would inhale a far greater quantity of spores, and my personal figure may be unrealistically low since I work with fungal cultures and search for fungi in the woods, but few humans will ever inhale more than a few grams of spores in a lifetime. So very little causes so much suffering for those allergic to fungal proteins. For someone blessed with an allergy-free life, the spores of most fungi go in, stick to the airway linings, and cause no problems. They are not recognized by the immune system and are evicted from the lungs by the slow ascent of mucus. (Go on, clear your throat.) I said *most* spores, because if they belong to pathogens, the smug non-atopics will succumb to infection just as easily as the wheezers, sneezers, and itchers. Similarly, if the spores ferry toxins, non-atopics will fare no better than anyone else. But these caveats don't interfere with the fact that people like the bathtub mycologist I mentioned in chapter 1 are very fortunate. You won't find them having breathing difficulties unless they have spent a few years in an asbestos mine, or have smoked an inordinate number of cigarettes.[17]

To understand what makes a mold spore allergenic, we need to think about the purpose of a spore. Fungal spores are survival capsules for the genome carried in their chromosomes. Because the spore carries the entire genome of the parent fungus, it has the potential to regenerate a whole mycelium, and later, to broadcast its own cloud of spores. (Humans behave in the same fashion, copying their genes, but polluting the planet with an expanding colony of glabrous hominids rather than spores.) Spore production far exceeds spore survival, so any individual spore has as little chance of germinating and producing a new mycelium as an individual mycologist has of starring in his own prime-time television show. But though the chance of survival is small, it isn't zero, and every spore is assembled to increase its personal odds. To this end, the instructions for making a new fungus are packed into chromosomes; the chromosomes sit inside the spore nucleus, which lies in the spore cytoplasm, and all this miraculous living mush is surrounded by a cell wall.

Biochemically speaking, the wall that protects the fungal cell is an incomprehensible structure. We know that it contains fibrous components and that these are embedded in a gluey matrix, but mycologists have nothing but vague presumptions about how the cell manages to blend these ingredients into a seamless veneer. The mixture has been likened to the structure of fiberglass: glass wool constituting the fibrous part, plastic forming the matrix. Chitin is one of the fibrous components. It is made from chains of a special kind of sugar molecule that are complexed into strands called microfibrils. Glucose and other types of sugar compose other kinds of fibrils, among which, beta-glucans are often mentioned by indoor mold investigators (figure 3.2). Much of the weight of a spore is made from strands of these molecules twisted into single or triple helices, so a measure of the amount of beta-glucan obtained from

Figure 3.2
Molecular structure of fungal beta-glucan. (From D. H. Griffin, *Fungal Physiology*, 2nd edition, New York: Wiley-Liss, 1994. Reprinted with permission.)

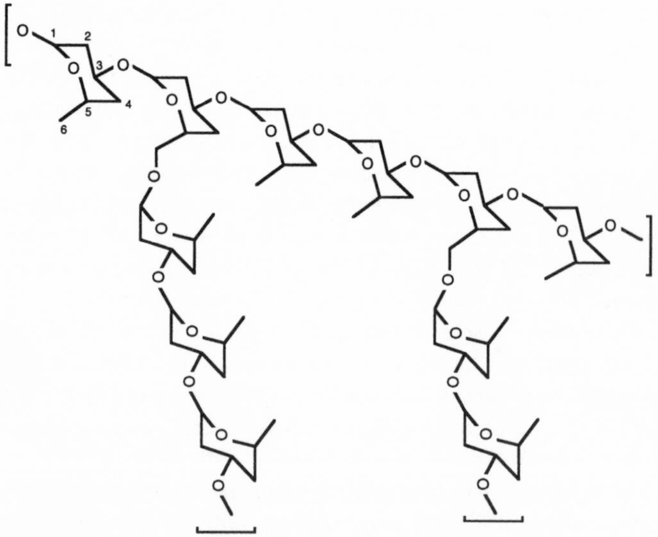

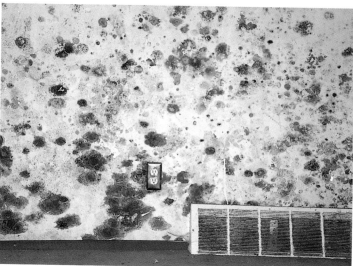

Plate 1
Interior of water-damaged home in Cincinnati severely contaminated with mold colonies. Photographs by Nicholas P. Money.

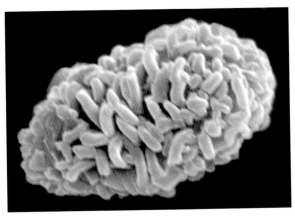

Plate 2
Highly magnified view of single *Stachybotrys* spore. Intertwined strands of unknown composition cover the surface. Young spores have a smooth surface. Photograph taken with scanning electron microscope by Matthew Duley.

Plate 3
Cluster of *Stachybotrys* spores at the tip of a single conidiophore. Photograph taken with scanning electron microscope by Matthew Duley.

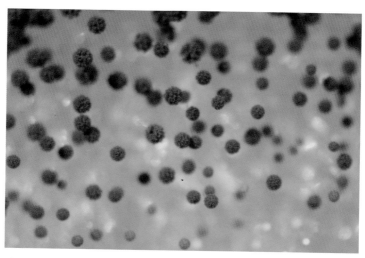

Plate 4
Globular masses of spores on the surface of a colony of *Aspergillus niger*.
Photograph by Nicholas P. Money.

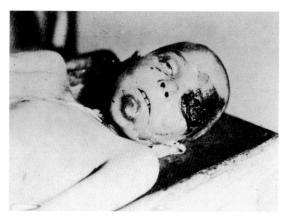

Plate 5
Severe necrosis of facial tissues of a child suffering from
alimentary toxic aleukia (ATA) thought to have been
caused by exposure to mycotoxins. From A. Z. Joffe,
"Alimentary Toxic Aleukia," in *Microbial Toxins. Volume
VII. Algal and Fungal Toxins*, edited by S. Kadis, A. Ciegler
and S. J. Ajl (New York: Academic Press, 1971), 139–189,
reprinted with permission.

Plate 6
Side view of *Stachybotrys* colony showing pair of conidiophores supporting clusters of spores embedded in
mucilage. Photograph by Nicholas P. Money.

Plate 7
Home in California damaged by dry rot.
Photographs courtesy of Luis De La Cruz (De La
Cruz Wood Preservation Services, Van Nuys, CA).

Plate 8
Rhizomorphs of the dry rot fungus *Meruliporia incrassata* entering home.
Photographs courtesy of Luis De La Cruz (De La Cruz Wood Preservation
Services, Van Nuys, CA).

Plate 9
Fruiting body of *Serpula lacrimans* showing rusty spore-producing surface. Photograph courtesy of Jim Worrall.

Plate 10
Wood decay in recently constructed home by unidentified wet rot fungus. Photograph courtesy of Luis De La Cruz (De La Cruz Wood Preservation Services, Van Nuys, CA).

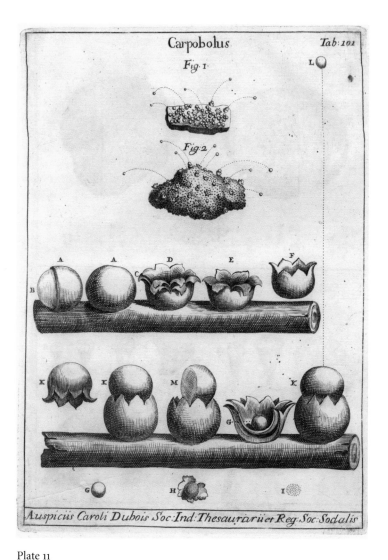

Plate 11
The earliest illustration of the artillery fungus, *Sphaerobolus stellatus* (referred to as *Carpobolus*), by Pier Antonio Micheli, in *Nova Plantarum Genera*, published in 1729. Courtesy of King Library Special Collections, Miami University.

an air sample is a good index of the prevalence of mold.[18] We encounter beta-glucans when we inhale spores and also when we eat mushrooms. Experiments on glucan inhalation suggest that they affect the inflammatory response, stimulating the production of T_H1 cells, which, as I said earlier, activate macrophages. Glucans also provoke macrophages directly by binding to receptor proteins on their surface, enabling the immune system to recognize fungi wherever they are encountered in the body.[19] Scientific interest in the immunological response to beta-glucans has been exploited by pharmaceutical companies to justify the sale of glucan capsules as a potent, nonprescription cure—unsullied by bothersome clinical trials—for every ailment linked to immune dysfunction:

> Every component of your amazing immune system is literally turned on SUPERCHARGED!
>
> —http://www.immunocorp.com (2003)

Immunocorp, in Irvine, California, obtains its beta-glucans from yeast, while other companies extract the polysaccharide from the fruiting bodies of bracket fungi like shiitake. Shiitake is one of the mainstays of herbal medicine, but evidence that orally administered glucans from this species, or any other fungus, produce any beneficial effects is very weak. I hope the stuff has marvelous properties, but call me a cynic. If invigorating one's immune system with high doses of fungal walls is shown to fight disease, I'll be the first to strip naked and dive into a tub of mushrooms.

If we want to understand which components in a mold spore have the greatest effects on the immune system, it seems more sensible to look at smaller molecules. Water-soluble molecules will dissolve from the intact spore into the airway mucus, and are less likely to be removed from the lung than fibrillar materials like wall glucans. A high proportion of individuals allergic to common molds possess IgE antibodies that recognize small fungal proteins.[20] In some cases, these allergenic molecules have been purified and identified as enzymes. Major allergenic proteins named Alt a 1 and Cla h 1, formed by the molds *Alternaria alternata* and *Cladosporium herbarum*, have been studied in detail, but their function in the fungal cell is not known. Sensitivity to these fungi is very common. There is a lot of interest in fungal proteases—fungal versions of the same kinds of allergens in mite excrement and pollen. In a screen of blood donations from the general population, researchers in Kansas City found that 9 percent of serum samples contained IgE molecules directed against *Stachybotrys* proteins, suggesting that sensitivity to the fungus may be very common.[21] A slew of proteases leach from *Stachybotrys* spores and are active against proteins called collagens that are major components of all of our solid tissues.[22] It is possible, then, that proteases in spores may provoke an allergic reaction in a similar way to the enzymes in mite feces: diving through the epithelium and into the underlying connective tissue to anger dendritic cells. But a direct link between particular proteases and IgE molecules directed against *Stachybotrys* is no more than a guess at this point.

Exposure to *Stachybotrys* may also induce a second kind of immune-response involving IgG. Virus-specific gamma globulins are injected into one's hindquarters to fend off hepatitis. But when large quantities of a particular antigen recognized by an IgG are inhaled, an illness called hypersensitivity pneumonitis (also known as extrinsic allergic alveolitis) can result. Under these conditions, clumps of antigens and antibodies called immune complexes develop. Degranulation of mast cells occurs at these sites, causing the local inflammation and increase in blood vessel permeability that should sound familiar if you survived my explanation of asthma. Fluid and cells called polymorphonuclear leukocytes—cells that function in pathogen destruction—escape from the leaky blood vessels and congregate around the immune complexes. If the inflammation continues, lung function is diminished by the accumulating junk. The commonest form of the disease is known as farmer's lung, which is caused by the inhalation of large numbers of fungal spores or bacteria from hay, grain, or even from a garden compost heap.[23] Again, soluble antigens leaching from microbes are the enemy. Hypersensitivity pneumonitis can also be caused by inhalation of dust from variety of agricultural products, wood bark, animal dander, and synthetic chemicals. Besides farmer's lung, the term encompasses sugar cane disease, cotton worker's disease (byssinosis), cork worker's lung, hot tub lung, mushroom worker's lung, bird fancier's disease, industrial bronchitis, and machine operator's lung. (I'm confident that "mold remediator's lung" will soon be added to the list.) Unlike asthma, which can occur immediately after exposure to the allergen, symptoms of pneumonitis usually develop hours after inhalation of the irritant. The symptoms include fever, coughing, and rales (crackling sounds heard with a stethoscope). Corticosteroids are prescribed to reduce inflammation, and future problems can be alleviated by avoiding contact with the allergens. For a farmer this may mean wearing a face mask in the barn, or getting someone else to move the hay. In a chronic form of the illness, the patient suffers from continuous shortness of breath, and coughing results from any exertion. Permanent lung damage may occur if the problem persists.

Claims have been made that hypersensitivity pneumonitis is associated with exposure to indoor molds including *Stachybotrys*, but, like the link with pulmonary hemorrhage, this remains unproven. Having said this, it is interesting that half the blood samples examined in Kansas City tested positive for *Stachybotrys*-specific IgG antibodies, and that these antibodies are also common among Scandinavians.[24] Whether or not this mold causes allergies, the prevalence of antibodies against *Stachybotrys* suggests that its proteins are finding their way into our tissues. This is surprising (if the research is solid) for a mold that was once supposed to be fairly scarce.

Proteins on the surface of a dead spore are just as allergenic as a protein on a live spore. Even fragments of spores or other components of a fungal colony can cause mast cell degranulation. But a living microorganism could certainly become a greater problem for someone with allergies if it were able to grow in the nasal passages or lungs. According to recent studies, *Stachybotrys* might be capable of doing this.[25] Since the Cleveland outbreak of lung bleeding mentioned in chapter 1, efforts have been made to develop animal models for testing the mold's behavior. If spore inhalation can induce lung bleeding in baby humans, investigators reasoned that *Stachybotrys* would do the same in baby mice, rats, or dolphins for that matter. (Because experiments on dolphins are poorly supported by the public, rodents secured this contract.) The experiments began with anesthetizing rat pups. Once subdued, the throats of the animals were nicked open and a catheter was used to squirt masses of spores into the trachea. To study the effects of the spores, rats were exsanguinated (bled to death) under anesthetic,[26] and their lungs were scrutinized. Not the nicest way to treat any animal, but nobody ever claimed that humans are, as a rule, nice to animals. (Like most Americans, I am opposed to vivisection, with the caveat that if it is necessary to slay a flock of bunnies with a lawnmower to keep my carcass running for an extra decade, then ladies and gentlemen start your engines.) The squirted spores found their way down into the rat lungs and, in addition to inducing an inflammatory response, caused damage to the lung tissue consistent with collagen breakdown. The tissue may have been damaged by the diffusion of those protease enzymes from the spores. More surprising was the discovery that the *Stachybotrys* conidia germinated, and, in some instances, the mold's hyphae penetrated the airway tissues. We're not talking deep penetration, just mycelial growth in the slushy contents of the lungs with some exploration of the epithelium. The rat experiments do not portray *Stachybotrys* as a spirited flesh-eater, but the fact that some of its spores germinate and hang around in the lung adds to concern about its proficiency as an irritant and toxin-producer.[27]

Spore size is a critical determinant of the movement of the fungus within lungs. When hydrated, each *Stachybotrys* spore is 7 to 12 micrometers (millionths of a meter) in length and 4 to 6 micrometers wide. This is quite large compared with other common indoor molds: *Aspergillus* species produce spherical spores with a diameter no greater than 5 micrometers. Some experiments show that only particles smaller than 1 micrometer will reach the finest ducts and smallest pockets in the lungs, the bronchioles and alveoli, but other studies have demonstrated that much larger fungal spores make the same journey with ease.[28] *Stachybotrys* spores are unlikely to penetrate very deeply when they are fully hydrated. Most will be carried out in the stream of mucus moving from the lungs and blasted toward someone else by a cough. But when the spores dry on walls and ceilings, and as they move through the air, they may shrink below the threshold for deep inhalation. This is a controversial issue that requires additional investigation.

Those convinced that serious home contamination by *Stachybotrys* and other molds is a recent phenomenon have suggested that this may be one factor in the increasing prevalence of allergies.[29] It is difficult to evaluate this idea because the symptoms of mold exposure are masked by the reactivity of our lungs to a legion of other biological and nonbiological materials: Atopics will feel the same whether mold spores, dust mites, or pollen made them sneeze or wheeze. Even when a mold is the source of the allergen, it is difficult to determine which mold, or molds, caused the allergic cascade. Consider a patient who lives in a home heavily contaminated by *Aspergillus*, *Cladosporium*, and *Stachybotrys*. He or she has developed severe allergies and tests positive for antibodies to one or more of these fungi. One would be forgiven for presuming that this is a clear instance of cause and effect, but the evidence is, almost certainly, questionable. Antibody specificity is a confounding factor that has often been overlooked, leading to the publication of faulty studies that purport to demonstrate allergy to particular molds. (An IgE that recognizes proteins extracted from a culture of *Stachybotrys*, for example, is likely to cross-react with proteins generated by other fungi.) Another problem relates to our inadequate picture of the varied mechanisms of mold sensitivity. An outbreak of respiratory illnesses reported among workers in a military hospital in Finland illustrates this.[30] The hospital had been flooded repeatedly since 1965, and many rooms were heavily contaminated with fungi. The most abundant was *Sporobolomyces salmonicolor*, a pink-colored yeast that jettisons its spores into the air using the same mechanism as a mushroom. The workers developed asthma, allergic rhinitis, and conjunctivitis, but did not show positive reactions to the yeast in a standard skin prick test. This argued against IgE-mediated allergy, suggesting that some other novel process may account for the way that some people respond to spore inhalation.

I have just noticed that the inner rim of a plastic cup I've filled with water is dotted with black mold colonies. Being a stupendous nerd, I have used the hand lens on my desk to resolve each dot into a spidery colony. Molds are everywhere and will outlive us by an eternity. Despite ongoing uncertainties about allergies caused by specific molds, there is no doubt that fungal spores cause a lot of grief for atopics and their families. But it was the conveyance of potentially lethal toxins by *Stachybotrys*, rather than allergies to spores, that allowed indoor molds to enter mainstream consciousness. It is these mycotoxins that make the mold story the most provocative encounter between humans and fungi since potato blight resulted in the Irish famine.

Hobbes clearly proves, that every creature

Lives in a state of war by nature.

—Jonathan Swift, *On Poetry* (1733)

Mycological Warfare

Fungi never intended to inflame or poison humans. After all, black molds spent hundreds of millions of years rotting dead plants before the evolution of wheezing authors and bleeding babies. Their spores were designed for drifting in air, not for swimming in lung mucus, so problems with molds are rooted in our insistence on breathing rather than any microbial malevolence. Enzymes toted on the spore surface perform various tasks for the fungus when it grows in soil or plant tissues, but unfortunately, the same molecules can also provoke an allergic response when inhaled. Of greater concern are mold toxins that clog biochemical machinery and destroy cells. Again, evolution didn't fashion these chemicals to attack humans. The fungal arsenal of toxins is directed against rival microbes that compete for the mold's food. In addition to other fungi, bacteria, soil protozoa (or protists), and invertebrates survive by consuming plant and animal tissues when they fall to the ground. If a mold gets there first, anything it can do to dissuade other diners will benefit its own growth, and pay dividends when it's time for spore production. Because a human being is built from the same components as other life-forms, a toxin that damages the cell membrane of a soil amoeba, for example, will probably interfere with the cell membranes of a baby. According to the bleakest mycological viewpoint, black molds veil humanity in an invisible blanket of toxic spores. But science also supports an alternative position, one that suggests that despite their toxins, black molds are an inevitably allergenic but otherwise harmless part of the living pollution that swirls around the globe. Since we sip from this soup every few seconds, we had better figure out which of these divergent views is correct.

Fungi produce hundreds of different kinds of toxins, a few of proven harm to humans, most of dubious significance. Everyone is familiar with poisonous mushrooms. A single fruiting body of a death cap or a destroying angel contains a few thousandths of a gram of amatoxins, which is adequate to kill an adult. Amatoxins are peptides that block protein synthesis in cells. The reason that the mushroom accumulates these poisonous molecules is unclear, but the most satisfying answer is that they defend the fruiting body against insect grubs.[1] Uncertainty also attends the function of other kinds of mushroom toxins, even though their virulence against humans is evident. Toxins generated by mycelia and spores rather than fruiting bodies are referred to as mycotoxins.[2] Although these compounds have an undeniably poisonous effect on cultured cells and laboratory animals, the consequences of ingestion and inhalation by humans are unclear because the dose is usually so small. Aflatoxins are produced by certain species of *Aspergillus* that grow on plant surfaces. These toxins make their way into us via contaminated peanuts, and dairy products and meat from animals that consume contaminated feedstuffs. Aflatoxins are formidable carcinogens, but nobody is sure whether we are exposed to enough of them to pose an appreciable danger. The situation is even more complicated when we try to assess the problems caused by the inhalation of spores that carry mycotoxins, but this is the issue faced by scientists making sense of the indoor mold crisis. In a severely mold-damaged home, there is no doubt that the occupants inhale many more spores than they would in a pristine living space. If *Stachybotrys chartarum* is identified, all hell breaks loose because mycologists have shown that this mold manufactures some very nasty substances. But what's the evidence that this particular black mold poses a danger for anyone lodging with it?

In his doctoral study of indoor molds, Kristian Fog Nielsen of the Technical University of Denmark found that *Stachybotrys* produced more mycotoxins than any other mold species he had examined.[3] Some of its mycotoxins rank among the most unpleasant in the catalog of toxic fungal metabolites, including trichothecenes, spirocyclic drimanes, triprenyl phenol metabolites, and stachylysin. None of these are sold on the Internet as a cure for the loss of memory, hair, or sexual performance, which is a sign that they should be avoided (because every other chemical seems to have its advocates in cyberspace—type "cyanide" or "battery acid" into your search engine and you'll find some nutcase selling them as guaranteed salves for recovering youthfulness). There is considerable variability in toxin synthesis between different strains of *Stachybotrys*, and genetic evidence is pointing toward trichothecene synthesis as an attribute of a specific group of strains that may soon be classified as a distinct species.[4] The absence of trichothecenes doesn't mean that the mold is benign, however, because other mycotoxins are generated by the trichothecene-free strains.

Trichothecenes were the first to be implicated in infant lung bleeding in Cleveland, and are a good place to begin this stocktaking exercise. During synthesis of these toxins, the ends of a chain of carbon atoms, interspersed by a few oxygen atoms, are closed by an enzyme to form two or more rings (figure 4.1). These compounds were discovered in the fungus *Trichothecium roseum* and turned up later in *Stachybotrys* and other molds. *Stachybotrys* specializes in the production of some of the larger molecules in this class, exceptionally toxic compounds known as macrocyclic trichothecenes. Connections between mold contamination of foodstuffs and illnesses in farm animals have probably been suspected throughout history, but the direct link between symptoms and trichothecenes was not established until the second half of the twentieth century. In chapter 1, I mentioned stachybotryotoxicosis in Ukrainian horses in the 1930s. Initially, the cause of the disorder was a mystery, so Soviet scientists referred to the condition as NZ, an abbreviation for *neizvestnoe zabolevanie,* "illness of unknown cause" (as graceless a term as stachybotryotoxicosis). Outbreaks of the disease continued during the 1930s in Ukraine and elsewhere in the Soviet Union and Eastern Europe, killing thousands of horses. Once the extent of the problem was recognized, it was redesignated as MZ, or *massovie zabolivanie,* meaning "mass illness."

The economic impact of an epidemic of equine stachybotry-

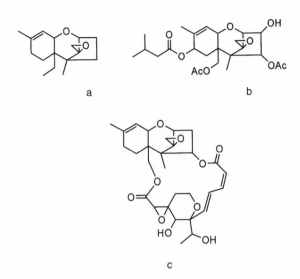

a

b

c

Figure 4.1
Chemical structure of trichothecenes. (a) Trichothecene nucleus, common to all of the trichothecene toxins. (b) T-2, a simple, non-macrocyclic trichothecene. (c) The macrocyclic trichothecene satratoxin G, produced by *Stachybotrys chartarum*. (Images courtesy of Daren Brown.)

otoxicosis would be almost imperceptible in most countries today, but at a time when horses were vital players in agriculture and transportation, research on MZ was accorded high priority. The All-Union Scientific Research Laboratory for the Study of Poisonous Fungi was founded in Moscow—an institution that did not survive the subsequent history of the Soviet Union—and many existing laboratories in the Soviet bloc became redirected to the study of toxic and pathogenic fungi. After the initial outbreak of the illness among horses in Ukraine, a variety of causes were considered, including toxic bacteria, plants, and insects. Efforts to understand the problem were further muddied by the proposal that mold contamination actually improved the nutritional value of animal feed. It is true that fungi can increase the protein content of plant matter and add some vitamins and minerals, but this is of little benefit if they also generate toxins. *Stachybotrys* entered the picture by name in 1940s, but its toxins were not identified for another 30 years. Soviet scientists reasoned that they were dealing with a toxic strain of the mold, known previously as a benign fungus that grew on plant debris, and named it *Stachybotrys alternans* var. *jateli*.[5] In the 1940s, researchers mimicked the poisoning symptoms by feeding horses with straw that had been deliberately inoculated with the mold. Taking this one step further, they were also able to poison horses by feeding them chunks of pure cultures mixed with oats. The severity of the illness depended on the quantity of mold culture given to each horse. A horse that endured a daily diet that included the contents of 10 Petri dishes died after a month, while a 30-dish dose proved lethal after five days.

There is a spectacular tradition of vivisection among *Stachybotrys* investigators. Here's a list of the animals who have made the ultimate sacrifice for black mold research: horses (as mentioned previously, and quite a lot of them), calves, sheep, and pigs (all force-fed because they were smart enough to refuse moldy straw), mice (Swiss mice and gray mice), guinea pigs, rabbits, chickens, dogs, ducklings, geese, protozoans, sea urchins, brine shrimp, insects, fish, frogs, and vervet monkeys. (A full set of references to this bloodbath is available from the author upon request.) The death throes of a bison and a hippopotamus also appear in the literature, but these animals were poisoned by eating moldy fodder that was naturally contaminated. Joseph Forgacs, a researcher in New York State, may have been the most zealous of the early *Stachybotrys* investigators. Here are a couple of quotes from his seminal review on stachybotryotoxicosis: "Two days after the first day of feeding, the mice developed atony [loss of muscle tone] and slight depression."[6] (I suppose Forgacs' diagnosis was spot on: Anyone would begin to dwell on the pointlessness of life in a little cage with nothing to eat but mold-contaminated oats.) He continued, "A pig weighing 125 lb. was force-fed a single dose of approximately 3.5 oz of the toxic straw and died 12 hours later."

Forgacs goes on to note that Soviet scientists learned that stachybotryotoxicosis follows two paths in horses: typical and atypical. In the typical form, poisoning is evident from the cracking of skin at the corners of the mouth and swelling of the lips to produce a hippopotamus-like appearance. While a horse with hippopotamus lips should be easy to spot, the poisoned hippopotamus mentioned above must have suffered invisibly. Inflammation of the nose and eyes can also occur, but that's all for a few days. Deeper indications of a problem become evident over the next couple of weeks, including a failure of the blood-clotting mechanism. Finally, the horse develops a fever, the number of leukocytes (white blood cells) crashes, vascular problems develop, pregnant mares abort their fetuses, and most of the animals die. This is the most common course of the illness. In the atypical cases, probably caused by the ingestion of massive amounts of toxic mold, the animal develops neurological symptoms. The horse loses its usual reflexes, becomes highly irritable, and is blinded. It stands with its legs wide apart or leans against something for support. Body temperature rises, and symptoms intensify with spasms and tremors and weakening of the pulse; death results from respiratory failure.

Pathological analysis of Ukrainian horses struck with either form of the disease revealed massive hemorrhaging of tissues with profound disruption of the mucus membranes throughout the digestive tract. The lungs were congested and often showed signs of hemorrhage, which is provocative given the symptoms of lung bleeding in babies seen in Cleveland more than half a century later. The damage to both the intestines and lungs is possibly explained by dual exposure through ingestion of moldy straw and inhalation of spores during feeding. Some Russian investigators claimed that stachybotryotoxicosis also involved tissue invasion by the fungus, after they had found that *Stachybotrys* proliferated from the cut surfaces of organs harvested from afflicted animals.[7] They confirmed that the fungus emerging from the organs was the same strain consumed by the horses, endorsing the idea that the animals were plagued by a deep-seated infection after consuming the toxin-producing mold. Forgacs was unable to find *Stachybotrys* in the tissues of any of the animals he poisoned, and even though the mold may grow for a while in the superficial layers of the lung lining,[8] the Russian claim of deep fungal invasion has never been substantiated. Unfortunately, the *Stachybotrys* cultures isolated by Russian researchers in the 1940s are unavailable today, raising the possibility that they were dealing with a different organism than the indoor mold linked to contemporary human illness.

Equine stachybotryotoxicosis persisted in Ukraine for about a decade, then subsided, though frequent outbreaks continued. Nikita Khrushchev, who was Commissar of Agriculture during World War II, has been credited for eradicating the mass poisonings by ensuring the supply of clean fodder for animals and ordering moldy feed destroyed.[9] Khrushchev's effectiveness in this position probably contributed to his subsequent political advancement.

Chemists at Stanford were first to suggest that trichothecenes could be the cause of stachybotryotoxicosis, and the connection was strengthened when these compounds were isolated from cultures of *Stachybotrys* in 1973.[10] Rather than killing horses with the purified toxins, the investigators confirmed the toxicity of the trichothecenes by murdering brine shrimp. In the arcane terminology of professional zoology, the brine shrimp is best described as a tiny lobstery thing that lives in saline lakes like the Great Salt Lake in Utah and Mono Lake in California.[11] The toxicity of many chemicals can be measured by exposing brine shrimp to increasing amounts of the substance and determining the critical concentration that slays half of the animals. This is an example of the LD_{50} test, which is more commonly undertaken by rodents and other animals closer to ourselves in evolutionary terms. The only problem with brine shrimp is that they are quite tolerant of toxins, requiring high doses to stop them monkeying around.

Photographs of dying sea monkeys are unlikely to leverage funds for research on *Stachybotrys*. It's better to cut to the chase and explain that when the toxins enter *our* cells they stop the synthesis of proteins. This is a problem because a human is mostly water, plus some minerals, and bucket or two of protein and fat: no water, dehydration; no minerals, no bones; no fat, no problem; no proteins, no Nicole Kidman.[12] No protein synthesis in those beautiful pink Australian lungs means respiratory failure. There is evidence that trichothecenes also cause chromosomal abnormalities, but this is probably a by-product of their effects on protein synthesis. Because they are such extraordinarily nasty compounds, tests for the presence of trichothecenes are crucial. A highly sensitive method is now available.[13] The test relies upon the reaction between a pair of molecules called luciferin and luciferase. These are responsible for the glow of fireflies and a host of other luminescent organisms including jellyfish, deepwater fish, and mushrooms. Luciferin is the incandescent partner but doesn't emit light until it reacts with luciferase, which is an enzyme. The test is a wonderful illustration of the power of molecular biology. It depends on the ability of the mycotoxin to halt luciferase synthesis, which serves as a flag for the inhibition of all protein synthesis. The test kit includes a mixture of luciferin, a mush prepared from rabbit blood cells, plus a scrap of genetic code that specifies the manufacture of luciferase. In the absence of the toxin, the rabbit mush behaves like an intact rabbit (or its cells), recognizes the instruction (which is a length of RNA), and translates this into the protein luciferase. This results in light emission. No light means no luciferase, which means trichothecenes. Very neat indeed. The luciferin assay is so sensitive that it can detect less than a billionth of 1 gram (a nanogram) of the toxin from a sample of indoor air.

Other methods to detect trichothecenes require expensive instrumentation. The mass spectrometer is a fabulously complex piece of hardware that can discriminate between the signatures of different organic compounds. It can show which of the 180 or so types of trichothecene are present, which is useful because some are far more toxic than others. (The luciferase assay does not have this capability.) For now, however, the technique is only useful if samples can be taken from a particular setting and analyzed in the lab, but there is great interest in miniaturizing the instrument so that it can be hooked to a portable computer and used for on-site screening of toxins. NASA's space station project may help turn this into a reality, because scientists working on the orbiting lab are under intense pressure to shrink scientific hardware, including spectrometers. Serious mold remediators would love to "look" in a moldy building and generate an immediate list of the toxins circulating in the air. It might even be possible to use this kind of technology to sniff out a black mold lurking inside a wall cavity without drilling any holes.

In addition to their effect upon protein synthesis, trichothecenes coax cells to commit suicide.[14] If a cell is injured and "bleeds" cytoplasm through its membrane, death is uncontrolled and doesn't involve the expression of specific genes. Cellular suicide, or apoptosis (pronounced *a-pop-toe-sis*, or *ape-a-toe-sis*), describes cell expiration when it takes place according to a genetically determined program. The regulated form of cell death turns out to be a vital one in human development: Apoptosis removes the tissue between our toes before birth, allows the inner lining of the uterus to become mobile in preparation for menstruation, and eliminates surplus cells as the brain is wired. It also participates in our response to viral infection and the normal functioning of the immune system, and allows the body to erase cells that sustain genetic damage.[15] The Nobel Prize in Physiology or Medicine for 2002 was awarded to Brenner, Sulston, and Horvitz, who pioneered this area of research.

In the study of trichothecene-induced apoptosis, investigators looked at the following roster of trichothecenes: satratoxin G, roridin A, verrucarin A, T-2 toxin, satratoxin F and H, nivalenol, and the aptly named vomitoxin.[16] This list was given in order of descending potency. Satratoxin G commands apoptosis even at low concentrations, and is one of the chemical calling cards delivered by *Stachybotrys* spores. Although other indoor molds produce some of the other mycotoxins on the list, none of them produce satratoxin G.

Enough of the trichothecenes for the moment; there are so many other nasties to consider. Spirocyclic drimanes such as stachybotrylactone and stachybotrylactam are complete and utter bastards of the highest caliber. Their mode of action is more subtle than the trichothecenes. Rather than killing cells directly or pitching them toward hara-kiri, they upset the immune system, leaving the victim vulnerable to bacterial infection. To explain this, we need to venture into immunology again to appreciate how healthy defenses throw off bacterial invaders. The liver produces a series of proteins that cooperate to destroy bacterial cells and signal that an invasion is in progress. These proteins constitute the complement system, so called because it complements antibody production. Complement proteins assemble on the surface of bacteria in the form of "attack complexes" that punch holes through their cell walls and membranes, causing them to ooze and die. They also mark bacteria for destruction by macrophages, and, after the proteins fragment, they disperse throughout the body where they serve as signals to summon other types of immune cell. By inhibiting the complement system, *Stachybotrys* might debilitate the defense against bacteria, increasing the likelihood of respiratory infections. Moreover, spirocyclic drimanes suppress the release from macrophages of a protein called tumor necrosis factor alpha (TNF-α). Normally, TNF-α increases blood vessel permeability, allowing fluid, proteins, and immune cells to infuse inflamed tissues; later, blood clots form in the blood vessels supplying the infected tissues, which serve to contain the infection. It is clear that any disruption of this vital mechanism will impair the body's ability to fight infection. Triprenyl phenol metabolites are precursor molecules that form during the synthesis of spirocyclic drimanes. They promote clot breakdown rather than formation, which may lead to the leakage of blood vessels in which clotting has already occurred. Did you think about the rich supply of blood vessels that envelop the lungs as you read the last sentence?

Finally, we come to stachylysin. Stachylysin is related to a larger class of toxins called hemolysins, which are produced by fungi and bacteria. One of these, beta-hemolysin, is secreted by certain strains of the bacterium *Streptococcus pyogenes*, including those responsible for necrotizing fasciitis or streptococcal gangrene (the infamous flesh-eating infection), and, more frequently, strep throat. Beta-hemolysins assemble mushroom-shaped complexes on the outside of cells and cause fatal membrane damage. In the lab, this activity is demonstrated with a simple but striking test. Sheep blood is mixed with agar to produce a bright red medium in a culture dish. When a culture of a bacterium that produces beta-hemolysins is grown on the medium, the blood cells surrounding the colonies burst, producing a translucent halo. In the human body, the effect of the toxins is similarly dramatic. By attacking the epithelial cells lining the lung and the surrounding blood vessels, beta-hemolysins can cause lung hemorrhaging in infants infected with *Streptococcus*.

Steve Vesper of the U.S. Environmental Protection Agency (EPA) in Cincinnati was the lead scientist in the discovery of hemolytic, cell-bursting proteins in *Stachybotrys*.[17] In this study, strains of the black mold from the United States, Canada, Europe, and Africa were cultured on pieces of sterilized wallboard, to mimic the growth conditions in water-damaged homes. After a few weeks of growth, spores were swabbed from the surface of the cultures and transferred to culture dishes containing sheep-blood agar. Within a week, cultures of strains taken from the homes in Cleveland in which infants had developed pulmonary hemorrhage showed the clear halos; after longer time periods, all of the 28 strains cultured on wallboard demonstrated hemolytic activity. Even strains isolated from homes in which infants remained healthy displayed this activity, though experiments indicated that these mold spores generated significantly lower concentrations of stachylysins. To probe the activity of stachylysins further, earthworms were recruited to the project. When the toxin was injected into the body segments surrounding the hearts—earthworms have five pairs of swollen blood vessels that function as pumps—they developed aneurisms in the major blood vessel that runs along their upper surface. The worms died within six hours of injection. This kind of vivisection I can stomach. The worms were purchased from a bait shop, so whether pierced with a syringe needle, or hooked by someone slouching in a plastic chair and breathing beer fumes, college wasn't in their future. Using antibodies raised against stachylysin, Vesper and colleagues have recently shown that the toxin is localized in the cell wall of the spores and mycelium of *Stachybotrys*, and also that the protein diffuses away from spores when they settle in mice and rat lungs.[18] They also claim that stachylysin can be identified in the blood serum of people exposed to *Stachybotrys*, which could lead to a highly specific clinical test for exposure to the mold.[19] Might this toxin be the smoking gun in human pulmonary hemorrhage?

Additional metabolites have been isolated from *Stachybotrys* cultures, including siderophores that are released from its cells and bind iron, compounds called atranones and dolabellanes whose function is unclear, and a type of cyclosporin that might suppress immune function if it is secreted in high enough quantities. The issue of quantity is an important one for assessing the significance of all of the putative toxins produced by molds. Investigators have estimated that a single spore of *Stachybotrys chartarum* can carry between 1 and 100 femtograms of macrocyclic trichothecenes, or a maximum of 0.000000000001 grams. If this is correct, the highly sensitive luciferase assay would detect the toxins carried by 10,000 spores. Harriet Burge, at the Harvard School of Public Health, used these numbers to assess the risk associated with spore inhalation.[20] She calculated that in a room saturated with mold spores, a person might accumulate a few tenths of a nanogram of toxin over the course of a day. This isn't much, but it is important to recognize that the local concentration of toxin right next to each spore could be sufficient to injure a small island of cells.[21] If thousands of spores reach the lung lining, they could act as thousands of foci of dead and dying cells; spread over the lungs, illness seems more likely.

Stachybotrys produces its toxins within the fluid cytoplasm of its cells. Once formed, the trichothecenes and other compounds are secreted and pass into the cell wall of the hyphae and the spores. If the function of these compounds is to dissuade other microbes from feeding around *Stachybotrys*, it makes most sense if they are released from the spores as they germinate on a fresh food base. This is the time when clearing the field of competitors is most important. Water may be available for a short time and if the mold is going to acquire enough food to produce the next generation of spores, it has to work fast. (The rapid flowering of desert plants following rain offers an excellent analogy.) When water hydrates the spore, the toxins will leach from its cell wall and diffuse into its immediate surroundings, casting a poisonous mist. The same thing happens on a damp wall and in the wet linings of the lung.

Toxin release is a passive process, so even when a spore is dead, it can still act as a source of poisons. Quite recently, mold researchers have begun to realize that spores are not the only vehicles for mycotoxin transmission. This was discovered by a team of researchers, including Tiina Reponen at the University of Cincinnati, who were studying spore dispersal from mold colonies growing on pieces of ceiling tile.[22] (The experiments proceeded from the studies discussed in chapter 2 that established the optimal airspeeds for spore release.) But in addition to the release of spores, the researchers found that clouds of minuscule particles were ejected under the same conditions. For every spore, more than 300 of these cell fragments were released from the tiles. The particles are no bigger than bacteria, but they are certainly capable of carrying mycotoxins. If future experiments validate this prediction, we may be able to explain how people can become sickened in rooms with moderate spore counts: Tiny particles could mist lung linings with mycotoxins even when very few spores are inhaled. If particle ejection occurs from colonies of *Stachybotrys* (experimental evidence for this has not been published), this would provide an efficient mechanism for mycotoxin delivery even if the larger spores were trapped in the upper airways.

Trichothecenes and the other toxins are examples of secondary metabolites. Primary metabolism enables cells to burn sugar and fat molecules to provide for their energy needs, to manufacture the components needed to sustain life and, where necessary, make new cells through cell division. Primary metabolites are the molecules involved in these essential reactions, and by and large biochemists know what these compounds do. Secondary metabolism describes other biochemical reactions whose function is less obvious. Black mold toxins fall into this category. Trichothecenes are highly toxic compounds, and it would seem reasonable to suggest that *Stachybotrys* and its friends make them to poison other organisms, along the lines I suggested at the beginning of this chapter. But we do not know this for sure. Mutants that fail to generate specific secondary metabolites can certainly survive in the laboratory, but this doesn't mean that these compounds are useless. Natural selection is too busy to carve out Byzantine biochemical pathways to enable something to synthesize a purple pigment, for example, if turning purple is of no value. Some authors have suggested that secondary metabolites are created as a way to detoxify compounds produced during primary metabolism, or as an energy sink when the cell has met all of its needs. Either of these suggestions are credible explanations for some of the things generated by cells. In the vast majority of instances, the products of secondary metabolism probably improve the fitness of the organism in some fashion.

Researchers dissecting the biochemical pathways that turn harmless precursor compounds into mycotoxins have found that multiple genes are involved, each encoding a different enzyme that catalyzes a step in the complex synthesis. In some cases, all of the genes necessary for production of a particular mycotoxin (or class of mycotoxins) are clustered together on the same chromosomes, forming what geneticists call super-genes.[23] Mycotoxin super-genes might be transferred as intact functional units between closely related species, introducing the complete biochemical apparatus for toxin production in a previously innocuous mold. This is called horizontal gene transfer. There are many questions about the way this works, but evidence that this has occurred is compelling. Genes that encode the synthesis of mycotoxins are genes that enhance the survival and reproduction of the organisms that carry them, making them genes that survive and reproduce.

In addition to its toxic secondary metabolites, *Stachybotrys* releases other molecules that might prove of clinical significance. Researchers are particularly interested in secreted proteases. I mentioned the proteases carried by *Stachybotrys* spores in the previous chapter, in relation to their possible role as allergens. The same enzymes might also be responsible for weakening lung tissue or damaging blood vessels: They certainly digest the collagens that account for much of the structural integrity of these tissues.[24] Researchers have also isolated a protease which they have named stachyrase A from a strain of *Stachybotrys* cultured from the home of a child suffering from pulmonary hemorrhage.[25] Though spores were washed from the lung of a 7-year-old boy in Houston suffering from a chronic cough and recurrent pneumonia, there is no direct evidence of *Stachybotrys* growth in human lungs. As I said in the previous chapter, however, the finding that it is capable of germinating in rat lungs and probing the lung lining with hyphae shows that this possibility cannot be discounted. Compared with a wet basement, healthy lungs are a place of formidable hostility toward a mold. Besides the mucus conveyor belt and immune cells bristling with chemical armaments, lung fluid contains proteins that inhibit the action of fungal proteases. A terrific battle of molecules is fought in the lungs. Some lung-dwelling microbes secrete another set of proteases that destroy the inhibitors: attack by the pathogen, counterattack by the lung, counter-counterattack by the pathogen! Isn't evolution beautiful? At least in a test tube, stachyrase A can snip and immobilize inhibitors produced by the lung, showing that the black mold possesses some of the attributes that permit survival in this unwelcoming environment.

Stachyrase A and other proteases are additional pieces of an immensely complicated puzzle. Instances of lung bleeding appear surrounded by so many guns with smoking barrels; at the moment it is impossible to figure out which weapon (or weapons)—which toxic metabolite or enzyme—might actually damage human tissues when carried by spores or smaller particles. But one thing is certain: *Stachybotrys chartarum* ranks among the most toxic microbes on the planet. Compared with other fungi, only aflatoxin-producing species of *Aspergillus* may be more harmful to humans (because we all encounter these cancer-causing compounds in our diets).[26] Other indoor molds generate toxins, but none such a varied catalog of poisonous compounds, nor in such high concentrations.[27] As I have explained, *Stachybotrys* poisons have the capacity to inhibit protein synthesis, interfere with the immune system, and make blood vessels leaky and cause bleeding. I say "have the capacity" to do these things, because experiments requiring the direct poisoning of human subjects have received little funding. Information on the toxicity of a trichothecene called diacetoxyscirpenol (DAS) or anguidine was obtained during clinical trials on the efficacy of a trichothecene in the treatment of cancer in the late 1970s and early 1980s.[28] The trials were discontinued because patients injected with anguidine suffered nausea, vomiting, diarrhea, and life-threatening loss of blood pressure. Severe skin irritation was confirmed as a symptom of human exposure to T-2 toxin—animal experiments had shown this before—when the purified toxins seeped inside the gloves of laboratory workers.[29] More information on T-2 came from the suicide attempt of a 28-year-old lab assistant who ingested milligram quantities of the purified compound.[30] The patient survived, and analysis of his blood and urine showed that the toxin was rapidly metabolized, supporting the idea that T-2 is more dangerous when it is absorbed through the skin (or from the lungs). The T-2 toxin is a trichothecene with a simpler chemical structure than satratoxin and the other macrocyclic compounds produced by *Stachybotrys*. Comparable information on the human toxicity of macrocyclic trichothecenes found in buildings isn't available, but their toxicity against cultures cells far exceeds that of T-2.

Traces of trichothecenes detected in "yellow rain" that fell in Laos in the 1970s led to the enduring claim that the Soviet Union had deployed mycotoxins against guerillas opposed to the Laotian communist government. Cases have also been made for trichothecene attacks by Egyptian or Soviet forces in Yemen in 1964; by the Soviet Union and its allies in Kampuchea and Afghanistan; and by Iraq during their war with Iran, within their own borders, and in Saudi Arabia. Government sources in the United States have estimated that these attacks killed 6,000 Laotians, 1,000 Kampucheans, and 3,000 Afghans.[31] The strength of the connection between trichothecenes and these casualties is thrown into dispute, however, by investigators who have claimed that yellow rain comes from the anuses of bees.[32]

The yellow rain story developed when Hmong refugees from Laos were interviewed by U.S. State Department officials in Thailand. They said that a sticky yellow substance had been sprayed on their villages and crops and had caused symptoms ranging from skin irritation and dizziness to skin blistering, seizures, and vomiting of blood. The majority of those interviewed said that many deaths had occurred in their villages. In a 1982 report to Congress, Secretary of State Alexander Haig announced that mycotoxins had been detected in samples of yellow rain collected in Kampuchea, charging the Soviet Union and its allies with breaking the 1972 Biological Weapons Convention and the earlier Geneva Protocol signed in 1925.[33] The alternative judgment against bees was based on the analysis of samples collected in Southeast Asia from the sites of alleged aerial attacks. Honeybee feces and the yellow rain samples comprised very similar mixtures of pollen grains from tropical trees common in the region, and entomologists testify that swarms of hundreds of thousands of bees do shower vegetation with yellow spots after leaving their nests. This explanation doesn't explain the identification of trichothecenes in yellow rain, but need not imply any fault on the part of chemists involved in the research, because strains of

Stachybotrys have been isolated from the feces of leaf-cutting bees and honeybees.[34]

Strong arguments can be made for and against the bee feces explanation, but the truth will not be known until the pertinent U.S. and Russian intelligence agency files are declassified. Means and motive are clear, particularly for the Soviet military. Equine stachybotryotoxicosis was guaranteed to stimulate some creative thinking in Russia, and I wonder whether the All-Union Scientific Research Laboratory for the Study of Poisonous Fungi was established in the 1940s with a purely peaceful mission. Soviet scientists were well aware at that time that trichothecenes were poisonous to humans. During the outbreak of stachybotryotoxicosis in horses, farm workers who handled contaminated straw and those who slept on mattresses stuffed with the same material also developed symptoms of trichothecene poisoning.[35] Secondly, a separate illness named alimentary toxic aleukia (ATA), or septic angina, afflicted 10 percent of the population of the Orenburg region of Russia between 1942 and 1947.[36] Near-famine conditions produced by the war impelled the population to consume unharvested grain that had passed the winter under snow cover. *Stachybotrys* wasn't to blame this time: The grain was heavily contaminated by a cold-tolerant, trichothecene-producing species of *Fusarium*. Symptoms of ATA began with hemorrhaging of the skin caused by fragile capillaries, and proceeded to widespread tissue damage with accompanying bacterial decay that created a repellent odor. Sickened individuals were disfigured by facial lesions that spread into the mouth and throat, and their vocal chords became overwhelmed with bacteria (plate 5). Hemorrhaging of internal organs was a common finding at autopsy. Not a pretty picture. Scientists working on this nightmare must have recognized that trichothecene-producing molds could prove a terrifying addition for their arsenal. I'm not saying that they acted on these thoughts, but a 10-second review of human history would suggest otherwise.

Acquisition of cultures for mycotoxin production and weaponization is straightforward. As I have said, the molds are ubiquitous and any mycologist could isolate cultures from plant debris or from water-damaged homes, and begin to screen them for toxins. Toxin-producing strains of *Fusarium*, *Stachybotrys*, and other molds with known characteristics can also be ordered from culture collections in the United States and Europe. They are listed in catalogs at Biosafety Level 1, the lowest, so they can be shipped to most countries with minimal paperwork.[37] Mycotoxin manufacture is relatively simple, providing that large-capacity fermenters, or bioreactors, are available. These equipment items are high-tech barrels and achieve the same result as a cellar full of the wooden ones in a traditional brewery or winery, by providing perfect conditions for the development of particular microorganisms. Indeed, many beers and wines are now fermented in stainless steel. The biotech industry also relies on fermenters to farm the genetically modified bacteria and fungi that produce pharmaceutical agents, food additives, and anything else we desire. (Artificial hair and teeth, I hope, having just frightened myself at the bathroom mirror.) The guts of the fermenter are sterilized, filled with nutrient broth, and inoculated with the microbe. Unlike the wooden barrel, the fermenter is programmed so that it maintains a specified temperature, level of dissolved oxygen and acidity, and stirs the brew at a constant speed to keep things mixed nicely. For some fermentations, solids such as rice grains are added to feed the microorganisms. A few days later, the cells are filtered from the broth, leaving liquid containing pharmaceuticals or bioweapons. Crude extracts of trichothecenes are yellow, which would be consistent with the color of the compounds found in the Southeast Asian villages. Experiments have shown that a few grams of crystalline product rich in trichothecenes can be extracted per liter of culture,[38] suggesting that a ton or more of the mycotoxin could be produced in a well-equipped fermentation facility. Purified trichothecenes are said to have a peppery odor.

Once the toxins are harvested, even in a very impure form, they could be dried and delivered as dusts, or dissolved in liquid and sprayed as droplets. Trichothecenes could be dispensed by aircraft, rockets, missiles, mines, or portable sprayers. According to military sources, trichothecenes act as blistering agents on exposed skin, mimicking the effects of acute radiation exposure. Unprotected ground troops would be incapacitated by severe skin and eye irritation, recalling the effects of mustard gas and lewisite in World War I. Animal experiments show that mycotoxins are far more lethal than these older chemical weapons: When droplets of trichothecenes are inhaled, the "targets" die within minutes.[39] For a soldier exposed to the toxins without protective clothing and face mask, treatment options are very limited. Skin contamination must be treated immediately by washing with soap and water, or use of standard issue decontamination kits. The U.S. Army's *Textbook of Military Medicine* says, "No specific therapy for trichothecene-induced mycotoxicosis is known."[40] This highlights the importance of protective clothing for military personnel (and the vulnerability of civilians to biological and chemical weapons).

In the late 1980s, Iraq manufactured mycotoxins in laboratories at Salman Park, south of Baghdad, which became their center for biological warfare research and development.[41] Investigations revealed that Iraq acquired fungal cultures from the Netherlands and purchased purified mycotoxins, including T-2 and macrocyclic trichothecenes, from a German subsidiary of the St. Louis-based Sigma Chemical Company (now Sigma-Aldrich).[42] These chemicals would be suitable for use as standards against which compounds purified from mold cultures could be compared in the lab. Although the extent of mycotoxin work in Iraq is uncertain, investigators established that Iraqi scientists produced 2,200 liters of an aflatoxin-containing solution and small quantities of trichothecenes. Though they are stupendously toxic, aflatoxins don't seem like ideal weapons because their cancer-causing effect isn't apparent to anyone at the time of exposure. Aflatoxins could serve as terror weapons against troops or civilians, but only if an alarming mycology seminar preceded the attack. It's also possible that aflatoxins have more immediate or acute effects on the human body when they are inhaled rather than ingested. A third possibility is that aflatoxins might be combined with trichothecenes or other toxins to concoct a mixture with a range of immediate and long-term effects.

The theory that Iraq had used mycotoxin weapons against coalition troops stationed in Saudi Arabia during the Gulf War in 1991 continues to circulate on the Internet, but supporting evidence is very slim. There is no question, however, about Saddam Hussein's motivation. For the Arab nationalist who allowed his fantasies of ruling a modern Mesopotamia to inspire the invasion of Iran and Kuwait, mycotoxins must have seemed an attractive weapon. They are relatively easy to produce, very cheap, and some of them are highly toxic. But did the Iraqis resume mycotoxin production once the arms inspectors withdrew? I asked Jonathan Tucker,[43] former director of the Center for Nonproliferation Studies and expert in chemical and biological weapons, for his opinion. He said that he thought it unlikely that evidence of weaponized trichothecenes will show up, but "given the major holes in our understanding of the Iraqi biological warfare program, anything is possible." Like the most paranoid government advisors, I thought about this a great deal in 2003 and convinced myself that the Iraqi military had hidden buckets of weapons-grade plutonium and enough anthrax to terminate our species. Measured against these nightmares, flasks of mycotoxins seem highly plausible.

And finds, with keen discriminating sight,

Black's not so black;—nor white so very white.

—George Canning (British Prime Minister 1827),

New Morality (1821)

Cleveland Revisited

Lung bleeding in infants is a serious condition. It can result from congenital malformations of the lung tissue, growth of benign tumors on the walls of blood vessels, infections by bacteria and fungi, and allergies to milk proteins.[1] It can also strike without an apparent cause. These cases of idiopathic pulmonary hemorrhage, or IPH, are very rare, afflicting one in a million babies. If an infant dies of IPH, the cause of death is usually recorded as an instance of the larger medical enigma called sudden infant death syndrome, or SIDS. Physicians have argued about SIDS for more than 200 years, ascribing the tragedy to asphyxiation (deliberate or otherwise), neurological, respiratory, and cardiac abnormalities, inflammatory reactions, low birth weight, secondhand tobacco smoke, viral infections, and a host of other causes. The number of SIDS cases has declined in recent years, due in part, perhaps, to the recommended practice of placing babies on their backs to avert suffocation in their blankets.[2] SIDS remains, however, the greatest cause of post-neonatal infant death, striking nearly 5,000 babies in the United States every year. The possibility that a parent may have harmed the baby cannot be ignored, and though this intensifies the agony for everyone involved, abuse is always investigated in a SIDS case. When investigators in Cleveland identified mold exposure as a potential cause of an IPH outbreak and began reexamining SIDS deaths that might have involved lung bleeding, they conceived an exceedingly controversial idea. Could indoor molds have been missed as a factor in some infant deaths classified as instances of SIDS?

When I began researching this book, I entered the toxic mold fray as a skeptic. America had embarked on a media circus that was vilifying fungi; as a mycologist, I had intended to repaint *Stachybotrys* as a harmless planet-mate. However, as I traveled from coast to coast, border to border, listening to physicians, attorneys, industrial hygienists, and fellow mycologists, and reading every relevant and irrelevant publication I could find, my belief in this mold's innocence faltered. *Stachybotrys chartarum* can produce highly toxic spores, and the fact that exposure to this fungus can be harmful has been firmly established since Soviet scientists investigated the epidemic of stachybotryotoxicosis among horses and humans in the 1940s. I don't know for certain that it killed anyone in Cleveland—nobody does—but I wouldn't recommend living or working in a building heavily contaminated with its spores. *Stachybotrys* and other molds are flourishing in American homes. Even if they are not damaging our health, they have certainly subverted many people's sense of well-being. Let's revisit Cleveland.

Do you know Cleveland? Located along 100 miles of the Lake Erie shore, Cleveland was long chastised for its burning river in 1969; no longer choked by oil and debris, the Cuyahoga has been "fireproof" for decades and is considered a "recovering" ecosystem. Cleveland has recast itself as "The New American City." The metropolitan area of 2.6 million people boasts world-class museums and a symphony orchestra, fabulous restaurants, and the Rock and Roll Hall of Fame. Face-lifted, fur-coated patrons of the arts[3] tend to live in idyllic leafy suburbs and gated communities offering private lakefront views, while across the tracks east of downtown Cleveland lie disheartening zip codes darkened by vandalized factories and warehouses. Surprisingly few residents of this less attractive district hold season tickets to the Cleveland Orchestra, and they are more likely to be poor and black than rich and white. More significant than its undeniable effect upon symphony attendance is the relationship between one's melanization and access to healthcare. In a March 2000 article on the racial health gap, Dave Davis and Elizabeth Marchak, reporting for the *Plain Dealer*, wrote, "In parts of Cuyahoga County, namely Cleveland's poorest neighborhoods, black babies are dying at rates that would embarrass some Third World Countries."[4] According to one version of the story I'm about to tell, *Stachybotrys* emerged as a source of additional suffering.

Rainbow Babies and Children's Hospital is part of the University Hospitals of Cleveland, affiliated with Case Western. Rainbow is ranked among the top 10 children's hospitals in the nation. In January 1993, Dorr Dearborn, a chief physician at Rainbow's Division of Pediatric Pulmonary Disease, treated a 22-month-old baby boy admitted to intensive care with lung bleeding. In keeping with the rarity of IPH, Dearborn didn't see another case for months. In October, a baby girl was admitted with the same condition. Two IPH babies in a single year was unusual, but no cause for alarm. But the girl suffered four more episodes of bleeding over the next few weeks, and in the following year another nine babies with bleeding lungs were treated at Rainbow.[5] Some of the patients appeared to have been treated successfully, but were readmitted after they began bleeding again at home. All of the patients were black, all but one was a boy, and they all lived in the same general area in northeast Cleveland. A 7-week-old boy died after his third episode of bleeding. The most intense symptoms were seen in a 9-month-old girl admitted in December 1994. She didn't show lung bleeding initially but was evaluated for rapid breathing (tachypnea), bluish skin color (cyanosis, caused by insufficient oxygenation of the blood), and grunting. Other than a brief nosebleed before hospital admission, there were no other symptoms of bleeding. But within a few hours of hospitalization, the girl suffered massive hemorrhaging of her lungs, resulting in respiratory failure. She died a few hours later. Something extraordinary was happening.

Lung bleeding can occur as an isolated event or persist in the form of intermittent leakages. One of the signatures of a slow trickle of blood into the lungs is the accumulation of iron within cells called macrophages. Macrophages crawl around in the mucus and acquire iron when they engulf any trespassing red blood cells or erythrocytes (which are filled with iron-containing hemoglobin molecules). Macrophages can be collected from the lungs by bronchoalveolar lavage, or lung washing, which involves the passage of a tube into one of the lungs. The same cells can also be seen in lung tissues removed for biopsy. Macrophages that have processed the iron from erythrocytes stain bright blue when treated with potassium ferrocyanide and hydrochloric acid (the Prussian blue reaction), providing a microscopic test for the condition even when the patient has no obvious symptoms. Pulmonary hemosiderosis refers to this combination of lung bleeding and iron accumulation.

Pulmonary hemorrhage accompanied by hemosiderosis was found in all of the Cleveland cases and was reported as a diagnostic feature.[6] Following the publication of the first paper discussing the cluster of IPH cases in *Pediatrics*, researchers from New Zealand wrote a letter to the journal editor to say that they had detected iron-laden macrophages in the lungs of infants who had been suffocated repeatedly over time.[7] (The murderers of these babies had the mental disorder known as Munchausen by proxy syndrome, in which harm is inflicted on the child to mimic a disease state and gain sympathy and attention for the child's caretaker.) The New Zealanders were concerned that child abuse might have been overlooked as the cause of the IPH cases in Ohio. The physicians and nursing staff at Rainbow had, of course, considered this explanation and where necessary had interviewed the parents according to standard procedure. The New Zealanders' suggestion was further eliminated by two arguments. First, the initial cases in Cleveland were clustered in one area of the city and occurred during a two-year period—it seemed unlikely that an epidemic of infanticide began in 1993. Second, the extent of the lung bleeding was far greater than that caused by suffocation—many of the babies required blood transfusions. Thanks to a rich archive of strangulation research by forensic scientists stretching back to the nineteenth century, we know how much bleeding occurs in cases of suffocation. In the 1890s, French pathologist Paul Brouardel—presumably he was insane—placed "a window in the chest wall of dogs, then [asphyxiated] them by applying a mask of soft wax over the face."[8] Implanted windows or fistulas are used in animal experiments so that the investigators can view internal organs and, by opening the window, withdraw samples from the digestive tract, all without killing the animal (yet). More recently, rats and rabbits have served as the unfortunate participants in suffocation experiments. In any case, animal experiments show that asphyxiation produces blood spots or petechiae on the surface of internal organs, not the substantial blood flow seen in IPH cases. This wasn't child abuse, at least not by human hand.

Dearborn's concern intensified with each new admission, and even though the total number of patients remained small, he alerted the CDC to the emergence of this highly unusual batch of IPH cases. The CDC scientists, led by Ruth Etzel, arrived in Cleveland the next day. Ruth Etzel is an expert on indoor air pollution and had previously studied the effects of lead and mercury exposure, passive smoking, and the inhalation of fumes by firefighters who had battled the burning oil wells in Kuwait after the Gulf War in 1991. With no immediate explanation for the IPH outbreak—the team had no reason to suspect that mold spores might be involved—they began by looking for anything unusual about the infants that might have led to bleeding.[9] The approach is a standard one for public health research, called the case-control (or case-controlled) study. Such investigations compare the lives of healthy and unhealthy individuals in the hope of identifying something uniquely damaging that may have led to illness. In a successful case-control study, something more common to one group than the other is spotted and a statistical argument is made that associates this factor with the illness, or with its prevention. This is the science of epidemiology. By the 1950s, epidemiological studies had demonstrated clearly that people who smoked a lot of cigarettes were more likely to develop certain kinds of lung cancer than nonsmokers. This didn't prove that cigarette smoking caused lung cancer, but it did lead to a whole body of scientific analysis that allowed us to reach this conclusion.[10] We all know that a person who smokes will not necessarily develop cancer, just that his or her risk of contracting the disease is intensified by continuing to smoke. Similarly, nonsmoking is no guarantee of avoiding carcinoma of the lung, but it does make this fate less likely. I know this is all so obvious that you must be yawning and reaching for your cigarettes, but too often we talk about cause and effect when science can rarely achieve this degree of certainty. Returning to lung bleeding, the investigators quickly identified some interesting clues, but they didn't claim that they had the answer.

Though all of the initial cases were black infants, the fact that the patients were born to low-income families was viewed as a more important feature of the IPH outbreak than race. The level of insecticides in the homes was examined to exclude the possibility that efforts to exterminate cockroaches might have sickened the babies; this seemed a distinct possibility, because pesticides had previously been implicated in lung bleeding among children in Greece.[11] The investigators found no significant differences between the levels of pesticides in the homes of sick and healthy babies in Cleveland, nor any differences in the levels of pesticide residues in blood samples from the two groups. The next idea was that lung bleeding might have resulted from secondhand exposure to crack cocaine—an idea favored by some commentators unduly influenced by the purported prevalence of drug use in the district. Again, chemical analysis of blood samples eliminated illicit drugs as a cause of lung bleeding, but it did appear that the sickened infants were more likely to live in homes where cigarette smoking occurred. On average, the IPH babies tended to have a slightly lower than normal birth weight, which might be expected to intensify the effects of any toxins. But among numerous other variables examined, one seemed most significant to the investigators: All the IPH cases had lived in water-damaged homes.

Cleveland is a wet city, suffering the heavy snowfall associated with the infamous "lake effect" in winter months and being a humid place for much of the rest of the year. The summer of 1994 was spectacularly damp, with 4 inches of rain falling in a single day in August. Homes on low-lying streets flooded as the city sewage system became overwhelmed and waterways overflowed. The investigators reported that there had been little or no effort to clean up the water damage in the homes of the sick babies, and pools of water were left to stagnate in the basements of some residences. A couple of additional details supported the idea that the home environment may have been to blame. As I mentioned earlier, bleeding occurred in some infants after they had returned home, and in two instances, relatives of the babies living in the same homes also developed lung hemorrhages. As the investigation proceeded, more and more attention focused on the babies' homes. The *Pediatrics* paper mentioned earlier,[12] the first peer-reviewed publication on the IPH outbreak, concluded that "The results of this investigation . . . suggest that affected infants may have been exposed to contaminants in their homes. Epidemiologic clues, such as water damage . . . suggest that environmental risk factors may contribute to pulmonary hemorrhage." Investigators began looking closely at the black mold they found growing in the water-damaged homes after it was identified as *Stachybotrys*.[13]

In 1986, William Croft and colleagues suggested that *Stachybotrys* was responsible for "a variety of recurring maladies including cold and flu symptoms, sore throats, diarrhea, headaches, fatigue, dermatitis, intermittent focal alopecia and generalized malaise"[14] Croft had studied the occupants of a large brick home in Chicago that had numerous plumbing and roof leaks. His conclusions about *Stachybotrys* were founded on four observations: (1) The cause of illnesses suffered by the household were a mystery; (2) the house was heavily contaminated with the mold; (3) highly toxic macrocyclic trichothecenes were extracted from moldy ceiling boards and dust that lined air ducts; and (4) the patients' symptoms resolved after the home was decontaminated. There was no attempt to carry out a case-control study in Chicago, but Croft's paper had established a precedent. As soon as the Cleveland investigators learned that Croft's microbe was flourishing in the homes of their cases, they knew they had uncovered a potential explanation for the IPH outbreak. But the media didn't reserve judgment—they reported that the cause of lung bleeding in Cleveland had been discovered and indicted *Stachybotrys* as a baby-killer.[15]

As I've indicated elsewhere, the route from potential explanation to convincing answer is fraught with difficulties. Let's recap a few of the problems. A toxin-producing mold growing in a particular home may pose a significant danger, but only if the occupants and the toxin-carrying spores, or colony fragments, make contact. The dispatch of spores or smaller particulates from a contaminated surface into the air breathed by the occupants—a process called aerosolization—is critical, and for the sticky colonies of *Stachybotrys*, this means that the fungus must be disturbed. The next issue is concentration: How much toxin is carried by each airborne particle, and how many of them must be inhaled to cause a significant allergic reaction or tissue damage? Even when a specific strain is isolated and can be grown in culture, its toxicity in the building where it was found is extremely difficult to assess because this is affected by the composition of the material it colonized, along with the temperature, humidity, and other environmental factors it experienced as it grew. These ambiguities undermine attempts to measure the toxicity of mold-contaminated buildings and also discredit guidelines for tolerable levels of mold exposure.

With these warnings in mind, we can further assess some of the results of the Cleveland study. The investigators reported far higher numbers of mold spores in the homes of sickened versus healthy babies, citing a 40-fold increase in airborne spores and a threefold increase in spores on surfaces. *Stachybotrys chartarum* was one of many species, but it stood out as the species more prevalent in case homes than controls: Surface samples revealed a 3,000-fold difference![16] Though far from proven, the case favoring *Stachybotrys* as the cause of lung bleeding was strengthening with each new publication. Scientists and clinicians were excited by Dearborn and Etzel's work, recognizing that a medical detective story was unfolding. The insurance industry was also absorbed by the events unfolding in Cleveland. If indoor molds were demonstrably poisonous, millions of Americans living in moldy homes might be at risk.

In 1999, someone spoke with a senior official at CDC, suggesting a detailed review of the work in Cleveland. I presume that this happened, because the CDC began questioning the work of its own field agent, Ruth Etzel. This was an extraordinary move at a time when further cases of lung bleeding were arriving at Rainbow Hospital and critical research was in progress. What reason did anyone have for thinking that something might be wrong? Had an insider criticized the work of the team in Cleveland? Or did someone else want the CDC to reach a particular conclusion? Remember, Dearborn and colleagues had not said anything declarative about *Stachybotrys* and IPH, only that this highly toxic mold might be involved. Now that I've made the suggestion that something conspiratorial may have happened, I also must suggest a plausible reason why I think this is the case. It's not difficult to do that here. The potential for insurance claims by homeowners maintaining that they were sickened by mold exposure is astronomical. The prospect of a peer-reviewed scientific article demonstrating a clear association between *Stachybotrys* and human illness must have kept plenty of insurance agents awake at night, and also made attorneys salivate. Also, everyone is aware that the insurance industry donates a great deal of money to finance political campaigns.[17] The CDC is a government agency. I presume, then, that somebody made a phone call from Washington, D.C., to Atlanta and told the CDC to shut those meddling scientists down in Ohio. Although this sounds like an episode of *Scooby-Doo* ("If it hadn't been for you interfering kids"), some conspiracies are real, even in the mysterious world of mushrooms, molds, and mycologists.

The CDC organized a panel of CDC scientists and outside experts to review the data collected in Cleveland. A preliminary report appeared in December 1999 and was followed by an unusual document published in the CDC's journal *Mortality and Morbidity Weekly Report*.[18] The articles list no authors, only the "Office of the Director, CDC." The panel concluded that the association between IPH and *Stachybotrys* was unproven. Few scientists familiar with epidemiological research would refute this statement, but Dearborn and colleagues were dismayed at the degree to which the CDC was attacking their studies. Here's one statement quoted from the report: "Panel members considered that the science in the CDC studies underlying these conclusions was 'flawed' and that the postulated associations should be considered, at best, not proven." An editorial note following the report stated, "Serious shortcomings in the collection, analysis, and reporting of data resulted in inflated measures of association and restricted interpretation of the reports." Journalists attempting to streamline the CDC's message offered the indefensible conclusion that mold had *not* caused the infant deaths.

When the quality of one's research is questioned in a peer-reviewed article, a scientist has the opportunity to respond to this by submitting a paper to the same or another journal. I did this once, following the publication of a paper by French authors that criticized my methodology in a study of fungal growth. It's clear to anyone reading my response that I disagreed with the independent appraisal of my experiments; other scientists who were fascinated by the details of fungal growth (probably all two of them) would be able to make up their own minds. This is the way that science is supposed to work. Dearborn, Etzel, and others in the Cleveland team were offered no opportunity to respond to their critics in the CDC's journal.[19] Because the CDC's evaluation was unsigned, nobody knew whether the critics were qualified to determine that the Cleveland work was unsound. This meant that other scientists grappling with indoor mold issues became aware that the CDC had dissed the original work but had no way of judging the merits of both sides of the dispute for themselves.

Since this is my book, I offer my penetrating insight. There were some faults with the original research, most resulting from the necessity of designing and conducting a case-control study within a very short time period. Ruth Etzel and her colleague Barbara Bowman arrived within 12 hours of Dearborn's contact with CDC; the fact that they were dealing with sick babies and an alarmed community accelerated the investigation beyond the normal pace of research. The key objections raised by the anonymous panel focused on the sampling methods used to determine the extent of mold contamination in the case homes. During sampling from the homes of IPH patients, the concentration of airborne spores was estimated by "pounding on furnace ducts several times, and walking on carpets."[20] They obtained higher spore counts in the homes of sickened babies than the control homes, but by deliberately disturbing mold-contaminated surfaces to "simulate household activities," they opened the investigation to serious criticism. Indoor mold experts now go to great lengths to disturb the living environment as little as possible to obtain an accurate measure of spore concentrations that reflect the exposure of occupants. There are limitations to this approach, because people certainly increase spore counts during their normal activities. But it is very difficult to replicate this disturbance scientifically. The CDC panel seemed overly interested in the possibility that crack cocaine had been involved in the lung bleeding, while the researchers made it clear that there was no evidence of drug abuse in the IPH households. They also made strong statements about the impossibility of spore inhalation deep into the alveoli without citing any definitive research on this question (I discussed uncertainties about this in chapter 3). Even if the spores do lodge above the alveoli, any toxins might diffuse through the mucus and rupture blood vessels clasping the airway endings. I'm not saying this does happen, only that it might, and that this is another idea that deserves further investigation. (The inhalation of minuscule cell fragments from mold colonies—as discussed in the previous chapter—wasn't considered by the CDC panel, because the importance of this mechanism has only come to light very recently.) I find it difficult to convince myself that the CDC panel offered an entirely objective assessment of the work in Cleveland. Why should anyone have expected that the relationship between molds and human health would be solved in a few months, especially when just a handful of babies had been sickened?

Since the release of the report from the CDC panel, other scientists commenting upon the research in Cleveland have faulted the investigators for failing to examine an eclectic list of factors that might—but very, very likely might not—have caused lung bleeding. Volatile organic compounds (VOCs) are often mentioned. VOCs arise from the building materials, household contents and residents, and from microorganisms (including indoor molds). Some VOCs have pleasant or unpleasant smells, others are odorless. Formaldehyde and other potentially harmful VOCs that "off-gas" from carpets are implicated in sick building syndrome,[21] but the health impact of most of these chemicals in buildings is unknown. Although the Cleveland investigators may not have studied VOCs in the water-damaged homes with sufficient enthusiasm to satisfy VOC specialists, Dearborn and colleagues did sample the air in the rooms for common VOCs and didn't find anything unusual.[22] There were many things, however, that they deliberately ignored. Raymond Harbison, Director of the Risk Analysis Center in Tampa, Florida, faulted Dearborn and colleagues for failing to "quantify confounding factors such as . . . the rate of formation of acids and salts from sulfur and nitrogen dioxide, the rate of formation of ozone, bacterial contamination, and dust mites."[23] Effective research never proceeds in this fashion: Scientists apply Occam's razor. Personalizing an ancient maxim, Franciscan friar and philosopher William of Occam (ca. 1285–1349) wrote (in Latin): Plurality should not be assumed unnecessarily. In science, this means that we don't look for all possible answers to a particular question, but focus on likely explanations first. If subsequent findings suggest that we were wrong, we then look at other possibilities. We don't reinvent the wheel each time we study a new problem. If the coffee maker doesn't work in the morning, a sane person doesn't suspect that a weasel is stuck in the filter basket: An animal might have fallen asleep in the machine, but it's more likely that the power chord is detached. In epidemiological research, the situation is a little different because the scientist is challenged to begin by looking at a range of possibilities, but that list is finite and is determined by the logic of the situation. Bacterial contamination is the only thing mentioned by Harbison that deserves additional study in future cases of lung bleeding. In bacterial infections, compounds called endotoxins can cause an overwhelming inflammatory response called septic shock. It is conceivable that the same compounds could be harmful in the indoor environment. I don't believe, however, that the investigators can be condemned for publishing their work on *Stachybotrys* before investing a decade exploring off-gassing from carpets!

Whether or not you accept that Dearborn and his colleagues furnished sufficient evidence to connect water damage, mold growth, and infant lung bleeding, it's worthwhile to look at studies by other researchers who became fascinated by the IPH outbreak. The mechanism linking fungal spores and hemorrhaging capillaries is very murky. A couple of questions are paramount. First, we must ask whether the spores produced in the Ohio homes carried any of the toxins discussed in the previous chapter. Bruce Jarvis, an expert on mycotoxins at the University of Maryland, compared the toxicity of nine strains of *Stachybotrys* isolated from case homes and nine strains from control homes.[24] Spores of these molds were used to inoculate culture flasks containing wet rice. After a four-week incubation, the moldy rice was ground in a coffee grinder and mixed with methanol to dissolve any toxins. Chemical analysis of the extracts showed that they contained trichothecenes and other mycotoxins; the biological activity of these compounds was confirmed by looking at their toxicity against cultured lung cells. But Jarvis found that not all of the strains collected from the case homes produced toxins, and that some of those taken from control homes did. In a later study, Steve Vesper, of the Environmental Protection Agency office in Cincinnati, examined air and dust samples collected from the home of an IPH patient. Though low concentrations of *Stachybotrys* spores were found in the air, a gram of dust contained a million or more of the black nuggets.[25] The spores were grown on sterilized pieces of cellulose-based wallboard in the laboratory for 10 to 30 days in an attempt to reproduce the conditions in the water-damaged home. Subsequent chemical study of these cultures produced an interesting result. The levels of macrocyclic trichothecenes—the toxins that the Cleveland investigators guessed might be responsible for lung bleeding—were very low, but the cultures did produce the blood-cell bursting proteins now called stachylysins. This may indicate that stachylysins are more important in lung bleeding than the trichothecenes, or that the most severe IPH cases are triggered by inhalation of a combination of mycotoxins.

It is important to recognize that Jarvis and Vesper didn't measure mycotoxin levels in the homes of the IPH babies. No mycotoxin-sensitive wand attached to a box with a red light has been invented, so everyone in the indoor mold field is forced to rely on less direct methods. These usually require culturing molds collected from the site, and there is no guarantee that the environmental conditions that prevailed in the home are faithfully reproduced in the laboratory. This is critical, because a particular mold strain may only produce high levels of toxins under certain conditions. Rather than solving anything, Jarvis and Vesper's analyses of the mycotoxins produced by the Cleveland molds have led to further questions about the IPH cases. This indicates—to me at least—how much work remains to be done, not that the original association between *Stachybotrys* and lung bleeding was fatally flawed.

Next question: If *Stachybotrys* spores cause lung bleeding, shouldn't we find them in the lungs of IPH patients? The fungus was isolated from the lungs of a 7-year-old boy with lung bleeding in Houston in 1999,[26] but while various bacteria and fungi were cultured from samples of lung fluid from some of the Cleveland babies, *Stachybotrys* did not appear. The presence of other organisms is not at all surprising: Our lungs are heavy-duty air filters. But the absence of the black mold is dumbfounding, unless those tiny fragments ejected from colonies are the real culprits.[27]

Significant issues about the relationship between *Stachybotrys* and lung bleeding are also raised by animal studies. Researchers in Finland studied the effects of introducing thousands or hundreds of thousands of spores into the nasal passages of mice. The spores induced an inflammatory response in the lungs, and hemorrhaging of alveoli occurred at higher dosages.[28] The Cleveland researchers also looked at the effects of *Stachybotrys* on newborn rats but introduced the spores into the wind pipes of the animals through a needle in an attempt to ensure spore inhalation deep into their developing lungs. We know that mice and rats are not perfect mimics for humans (with a few honorable exceptions), but the mold toxins strike at such fundamental aspects of cell function that we have good reason to expect that they would cause severe lung damage in rodents if *Stachybotrys* does cause IPH. Rat pups treated with the spores of a toxin-producing strain isolated from the home of an IPH patient showed changes in lung function, and analysis of lung fluid showed iron-containing macrophages and evidence of severe inflammation. Fatal lung hemorrhaging occurred in 83 percent of the treated animals, but only at dosages of millions of spores.[29] So although the symptoms of poisoning echo those of the IPH infants, it is hard to imagine how a baby—rat or human—could inhale these numbers of spores. (The effect of colony fragments has not been tested yet.) Research on the animal model for IPH continues.[30]

If no more babies had been sickened after the original cluster in 1993 and 1994, interest in the outbreak among physicians and researchers would surely have evaporated, but this did not happen. By January 2000, Dearborn and colleagues had treated a total of 30 babies at Rainbow suffering from IPH. Here are some statistics from Cleveland: 26 of the infants sustained major lung damage, four died from massive lung bleeding, one more infant died two months after treatment (and had also developed a viral infection in his lungs); 89 percent of the cases lived in water-damaged homes containing toxin-producing molds. The same percentage of the cases were regularly exposed to tobacco smoke, so the importance of secondhand smoke remains part of the investigation. When medical records from the entire Cleveland area are considered, the total number of IPH cases rises to 41, with 12 infant deaths. The number of cases in the rest of the country is ambiguous, but patients with a similar profile to the babies sickened in Cleveland have now been reported in Chicago, Houston, Kansas City, and Wilmington, Delaware. In a study of eight cases in Chicago, the babies were scattered throughout the city rather than confined to a particular district. In another departure from the Cleveland cases, samples of the lung fluid did not contain masses of iron-laden macrophages: Lung hemorrhage occurred, but little hemosiderosis.[31] A single case from Kansas City was interesting because most of the house was found to be relatively mold free.[32] But a roof leak had occurred above the room occupied by the 1-month-old boy, leading to the proliferation of numerous fungal species, including a toxin-producing strain of *Stachybotrys*. So low levels of other molds drifted throughout the house, but *Stachybotrys* was confined to the boy's bedroom.

Stachybotrys is also implicated in illnesses among adults in the United States and in Europe. Horticultural workers in Germany who handled paper plant pots covered with the fungus developed severe inflammatory reactions,[33] and numerous reports of building-related illness among office workers have been associated with the prevalence of *Stachybotrys* spores.[34] None of these studies entailed the kind of case-control investigation carried out in Cleveland, but together they make it difficult to dismiss the impression that conditions favoring the health of black molds are incompatible with human welfare. On the other hand, the strength of this claim is all too easily tarnished by studies that stretch one's credulity beyond breaking point. William Croft—senior author of the 1986 study in Chicago[35]—published an extraordinary paper in 2002 in which he identified trichothecenes in the urine of patients living in mold-contaminated buildings.[36] His patients came from Nevada, Wisconsin, Kentucky, and New York City, and displayed a frightening array of symptoms including "burning eyes, severe headaches, respiratory difficulty, depression, diarrhea, severe chest tightness, nasal burning, leg cramps, dental problems and skin rashes." Some of the patients manifested neurological impairment: "The young boys [from Kentucky] could not tie their shoes or ride their bicycles very well and endured overall stunted development." Others developed agoraphobia and other psychiatric problems, and life-threatening illnesses including skin cancer. This overview doesn't do justice to the mind-boggling range of illnesses that Croft believes were caused by mycotoxin exposure. He then reported that weanling rats injected with toxins extracted from urine samples taken from the patients developed severe tissue damage, including rupture of alveoli, and exhibited symptoms consistent with neurological damage.

Based on these results, Croft insisted that he had confirmed the connection between trichothecenes and human mycotoxicosis. When news of the publication circulated via E-mail between *Stachybotrys* researchers it caused a great deal of excitement, but its scientific validity is questionable at best. (Now, now! Don't start pointing at me and shouting hypocrite. I am the author of the previous sentence, not a nameless committee organized by the CDC, and I'll be happy to justify my critical opinion.) The chemical techniques used to identify the trichothecenes were alarmingly crude, and Croft's own urine was employed as the only control sample (which, he was happy to report, was unblemished by mycotoxins). Besides these problems, it is impossible to assess the relationship between molds and any of the symptoms catalogued in the paper, because the study lacked any hint of the kind of careful epidemiological investigation conducted by Dearborn and colleagues (whose work, bizarrely, Croft failed to mention).

Now, I can hear your brain-hamsters running faster in their metal wheels—doesn't this raise questions about the same author's 1986 study in Chicago? Yes and no. The strangeness of the new paper does make one doubtful about the causal link between *Stachybotrys* and the varied illnesses contracted by the Chicago household. But the real value of the 1986 study lies in the novelty of Croft's suggestion that there might be a causal relationship between the sicknesses of people living in a house contaminated by *Stachybotrys* and the mold that did the contaminating. Previous suspicions about stachybotryotoxicosis in humans had been limited to problems associated with contaminated animal feed. Croft's 1986 study was bereft of epidemiological finesse, but it did encourage Dearborn and colleagues to look at the toxic-mold hypothesis more carefully when faced with IPH patients from water-damaged homes. Scientific understanding often proceeds in this inductive fashion, each piece of research adding a little more to the picture. Like most scientists, Croft added less than he thought, but his contributions may still have been important. Only through continuing research on unexplained illnesses in mold-damaged homes will we reach a satisfactory conclusion about the toxicity of indoor molds.

Until we have a definitive answer, one that is blessed with widespread concurrence in the scientific community, we are faced with a dilemma. Should we dismiss the possible threat posed by *Stachybotrys*, and other fungi, in homes? The CDC hasn't gone this far. Instead, it occupies a holding pattern, maintaining its criticism of the Cleveland study but recommending that areas of mold growth in homes be cleaned and that "sensitive individuals" avoid exposure to spores. Official concern is limited to problems of mold allergy and fungal infections in people suffering from chronic illnesses.[37] The CDC's "Molds in the Environment" web site makes no mention of the special danger that *Stachybotrys* may present for infants.

Concern is unlikely without empathy, and the number of families impacted by IPH is so very small that public indifference is not surprising. After I gave a talk on mold biology at a convention, a man introduced himself and said he had been struck by my description of the horses dying in Ukraine. He and his wife had lost their adopted son within days of his birth in a city hospital. Shortly after collecting him from the neonatal unit, the baby began gurgling and coughing up blood. Despite aggressive medical treatment, the boy died. Abuse of the infant was investigated and rejected, and the cause of death was listed as SIDS. An autopsy showed that he had suffered lung bleeding and hemorrhaging of other organs. The father knew that there wasn't much I could say; none of the physicians who treated his baby had been able to explain the tragedy.

I don't know if *Stachybotrys* is guilty of killing babies, but I think its presence in the homes of IPH patients is more than pure coincidence. *Stachybotrys chartarum* is one of the most poisonous molds on the planet. It generates toxins that inhibit protein synthesis in animal cells, plus other metabolites that have the capacity to interfere with the immune system, make blood vessels leaky, and cause bleeding. Whether or not they have ever been deployed, these mycotoxins are nasty enough to be discussed as battlefield weapons. *Stachybotrys* certainly kills horses that consume contaminated fodder and has sickened people exposed to the same materials.[38] Animal studies show that its spores can damage lung tissue (admittedly at ridiculously high concentrations), and there is new evidence that it can grow in animal lungs.[39] Antibody profiles suggest that humans are exposed to *Stachybotrys* spores far more frequently than anyone has imagined.[40] Cleveland infants living in homes contaminated with *Stachybotrys* spores bled during or after exposure, and sometimes repeatedly with each reintroduction to the same moldy environment.

If you disagree with this viewpoint, ask yourself whether you would feel comfortable tucking your children into bed next to a heating vent spewing black spores. I'm not saying that *Stachybotrys* causes lung bleeding, but I wouldn't gamble anyone's health by saying that they shouldn't worry about mold inhalation. I'll conclude this chapter by suggesting that I believe researchers have established very reasonable doubt that the spores of this organism are harmless. Now we turn to a newly identified species, one that has formed a firm symbiotic relationship with *Stachybotrys*. This impressive beast is called the black mold attorney.

When I hear of "equity" in a case like this,

I am reminded of a blind man in a dark room—

looking for a black hat—which isn't there.

—Lord Bowen (British judge, 1835–1894)

Your Verdict, Please

At a meeting on mold litigation in San Antonio in 2002, speaker Everette Lee Herndon, Jr., woke me from a pleasant afternoon drool—warm notebook computers are such seductive desk pillows—with this phrase: "Stop the water damage, stop the hemorrhaging." It took me a few seconds to realize that he wasn't offering advice about lung bleeding. Herndon, a claims consultant and expert witness in mold-related lawsuits, was referring to the hemorrhaging of funds from insurance companies sued by homeowners.

The meeting had been organized by HarrisMartin, a firm that publishes a line of monthly journals on legal proceedings arising from the undesirable effects of vaccines, silica and asbestos, drugs and dietary supplements, and, of course, molds. A year earlier, a Texas jury had awarded $32 million to homeowner Melinda Ballard in settlement of her suit against Fire Insurance Exchange, a member of the giant Farmers Insurance Group. The jury held Farmers responsible for multiple failures in settling an insurance claim to rectify water damage that had led to subsequent mold damage to Ballard's home.[1] Other mold lawsuits had been fought, and a handful of colossal awards were made by juries around the time of Melinda's coup, but this judgment reinvented *Stachybotrys*—a microbe that had passed millions of years munching invisibly through incalculable masses of plant products—as a star of screen and page. The Ballard case is an obvious landmark in the toxic mold story, but to those battling for and against her claim, it wasn't really about fungi at all.

Melinda Ballard lived with her husband and son in an enormous house—7,400 square feet—in Dripping Springs (I'm not making this up), west of Austin, Texas. Modeled after Tara in *Gone with the Wind*, the house sits on a 73-acre property that includes a nanny's apartment and other outbuildings, swimming pool, pond, and tennis court. It was purchased for $275,000 in a 1990 foreclosure sale; understandably, Melinda thought she had a bargain. Following a cycle of plumbing leaks, a variety of molds became enchanted by the home and moved in with the Ballards.

In 1998, a leaky toilet flooded one of the bathrooms, and a hardwood floor in another part of the house began to warp. The toilet was repaired and sodden floor boards were replaced, but the water damage kept spreading. A leaking refrigerator added to the mess, and although high moisture levels around the refrigerator were detected by a civil engineer employed by the Ballard's insurers, the leak wasn't fixed. Farmers paid for the early home repairs, including more than $100,000 to replace the flooring. Worth far more than the purchase price, the home had been underinsured, which complicated the settlement of the claims by Farmers. There was no mention of *Stachybotrys* at this time, though the molds were having a feast. This changed with a chance encounter in April 1999, when Melinda met an indoor air quality consultant called Bill Holder, during an airline flight. She recounted the story of her damaged home and said that her husband, Ronald Allison, had developed memory problems since the water damage and now had trouble concentrating on the simplest of tasks. Her son began suffering from asthma symptoms, too. During the flight, Bill Holder introduced Melinda Ballard to the world of mycology; a few days later, he visited Dripping Springs and took air samples from the house. *Stachybotrys* was confirmed as one of the lodgers. The Ballards moved into the nanny's apartment, then, after further tests confirmed the extent of the mold growth in the house and its presence in the apartment, they left all of their belongings and checked into a hotel in Austin. They would never live in the house again.

Farmers continued to settle the multiple claims made by the homeowners: $8,000 for a shower leak, $25,000 for a leak in an ice-maker, and more than $45,000 for damage to walls and sheetrock. Time for a brief reality check. All of the plumbing fixtures in my home must have been manufactured in North Korea, because whenever I try to replace something, I'm greeted with the phrase "I've never seen anything quite like this before" at my local hardware store. My stepkids have twice flooded the house by plugging an upstairs toilet, and a thunderstorm of biblical proportions flooded our library in 2002. Nevertheless, I have managed to keep the molds at bay with relatively little hardship and a single $2,500 insurance claim, which makes me wonder what the plumbers responsible for the Ballard's pipes are doing now. (Not fixing toilets, I hope.) Rather than mailing Polaroids of the plumbers and a wad of cash to a gentleman listed in the "Hit Men" section of Yellow Pages, Melinda decided on a more profitable solution. She sued Farmers for "breach of contract, deceptive trade practices, breach of duty of good faith and fair dealing in the claims handling process, and negligence." After failed attempts at mediation, the case proceeded to trial in May 2001. Her attorney was Mike Duffey, a partner with Childress and Zdeb, in Chicago. Duffey is a courtroom silverback. Farmers Insurance Group was in deep trouble.

I already told you that Melinda was awarded $32 million by the Texas jury, but the breakdown of the number is interesting. In round figures, the jury said that Farmers should pay $2.5 million to replace the home, $1.2 million to remediate the water damage, $2 million to replace the Ballard's contaminated belongings, plus a few hundred thousand dollars for additional expenses incurred by the family. The remaining $26 million was split between $5 million for Melinda's mental anguish, $12 million in punitive damages, and $9 million in attorneys' fees. Given the size of the total award, Duffey and colleagues earned every penny.

Surprisingly, the scientific debate about *Stachybotrys* and its cocktail of poisons didn't play a major role in the trial. In awarding $5 million to Melinda for mental anguish, the jury empathized with her spirited account of the terror she felt once she learned that the home had been colonized by a toxic mold. But rather than ruling on the actual threat to the family's health, this decision rested on the argument that Melinda's anxiety would have been avoided if the insurers had honored their responsibilities toward the homeowners. While the impression of a home eaten away by a black fungus certainly added to the jury's evident displeasure with the insurer, most of the mycology was eliminated from the trial. Two expert witnesses—Wayne Gordon and Eckardt Johanning—had been asked to testify that Ronald's mental problems resulted from mold exposure.[2] The presiding judge, the Honorable John Dietz, excluded this personal injury part of the lawsuit because he was not convinced that sufficient scientific evidence supported a causal relationship between mold and brain damage. (During pretrial questioning, Wayne Gordon agreed with this appraisal, which isn't surprising given our fragmentary understanding of mold toxicity.) So the jury's record-breaking judgment was recommended because of the insurers' bad faith and fraud in their dealings with the Ballards, not because the conduct of Farmers's agents led to a life-threatening mold problem. For anyone but a lawyer, this may seem a subtle point. But it does indicate that members of the legal profession are well aware of the uncertainties about mold toxicity discussed in this book.

For most Americans who have watched television interviews with Melinda Ballard, or read articles about the case—and it would have been difficult to have missed it entirely—the specter of toxic mold overshadows insurance company malpractice. *Stachybotrys* is so much sexier than the small print in a homeowner's policy. Melinda has appeared on the news magazine *48 Hours* on CBS, seemed like a regular fixture on ABC's *20/20* for a while, and showed up on many other programs on television and radio. She is a producer's dream because she wields metaphors and one-liners better than a talk-show host. I'll provide an example. Speaking about homeowner's experience she said that they had "entered into the abyss" when the biohazard-suited hygienists showed up for a look-see, and that "Everything they've ever worked for, disintegrated before their eyes."[3] The print media was similarly enamored of the Ballards' story. "Beware: Toxic Mold" wrote Anita Hamilton in *Time* magazine.[4] The photographs in this article are very provocative. A sidebar that lists suggestions on protecting homes from mold damage is highlighted with the standard yellow symbol with the words "Infectious Substance." This is very misleading: *Stachybotrys* and most other indoor molds may be allergenic, and may carry toxins, but they don't infect human tissues. Other pictures in the same article depict a home inspector wearing a biohazard suit and face mask, and a mold-contaminated home being burned for lack of other options. The *New York Times Magazine* featured a well-crafted article by Lisa Belkin titled "Haunted by Mold"; the subtitle read, "It grows in the walls. It chokes your child and renders your husband senseless. It's your—and your insurers'—worst nightmare."[5] In this piece, the Ballards are pictured outside their home. Melinda's 5-year-old son is wearing a gas mask.

I'm certain that Melinda Ballard has slipped into the nightmares of a few insurance industry executives. "She is brassy, fast-talking, occasionally profane," wrote journalist Claudia Grisales.[6] I'd add "intelligent," "sharp-witted," and "aggressive" to the adjectives. Ballard describes the judgment against Farmers as a self-inflicted wound they would have avoided by treating her fairly. Since the publication and broadcast of stories about her fight with *Stachybotrys* and Farmers, the number of insurance claims related to mold damage have skyrocketed. Farmers dealt with 12 cases in Texas in 1999—before the verdict—rising to 12,000 in 2002. The Insurance Council of Texas estimates that mold claims since 1999 have exceeded $3 billion.[7] Farmers and other insurance companies have considered withdrawing from the homeowners insurance market,[8] and it is now almost impossible for anyone to obtain coverage for mold damage in Texas. As might be expected, some insurance executives admit to feeling a little discouraged by Melinda's opinion of their almost philanthropic record of public service. In the Claudia Grisales profile of Ballard,[9] Bill Miller, a spokesman for Farmers, was quoted as saying, "I think she hit the jackpot with the jury award on a fraudulent case." Miller continued, saying, "I don't believe anything she says about her case. . . . She has no credibility." Dangerous words. In December 2002, Melinda sued Farmers and Bill Miller and his company Hillco Partners for libel.

After the $32 million judgment, Melinda formed Policyholders of America, an organization that assists consumers who have experienced problems with their insurance companies (generating a limitless roster of potential clients), publicizing the mold crisis, and lobbying for reform of the insurance industry. I think I speak for all Americans when I say that we are delighted to pay a great deal for all kinds of insurance. We are, however, vulnerable to becoming mildly irritated if difficulties are encountered in obtaining compensation when a claim is filed, and can even succumb to feelings of apprehension if the insurance company then increases our premiums. Policyholders helps consumers to express these emotions on its web site, through a series of pop-up warnings to insurers: "We've got news for you, Jerky [presumably this label is directed toward an insurance company executive] . . . Your worst nightmare has just become a reality . . . We're united . . . There are hundreds of thousands of us. And we're pissed." Despite our love for insurance companies, many people are glad that Melinda Ballard is willing to fight. The fact that she's brassy, grew up in an affluent family, and drives a Jaguar (this is always mentioned by journalists) is horse feathers.

A few days before Christmas 2002, Melinda's personal triumph over the insurance industry was numbed by an appeals court ruling that slashed the $32 million award to a mere $4 million. The court found "insufficient evidence to support the jury's findings of unconscionability or fraud."[10] It upheld the jury's finding that the insurer had breached its duty of good faith and dealing toward Ballard, but reversed the jury's award for punitive and mental anguish damages because there was "no evidence . . . that [Farmers] 'knowingly' breached its duty." Unabashed, Melinda immediately announced that she would appeal the appeal. She still maintains that an alliance of hopeless plumbers, *Stachybotrys*, and her insurance company destroyed her home and disabled her husband (he is no longer able to work). There seems little hope for an agreeable conclusion to this story.

I want to address some of the wider implications of the Ballard case now. *Ballard v. Fire Insurance Exchange* was an exceptional lawsuit in the history of mold litigation, but the phenomenon of mold contamination affects everyone. Because the appearance of mold is viewed as a reason for concern, quite apart from the mess caused by leaking pipes, its presence makes it easier for an attorney to argue that the homeowner is dealing with a crisis. If water damage occurs without any mold growth, everyone might feel satisfied by drying out wet flooring and paying a reputable plumber to fix the leaks. But the threat of a personal injury claim for illnesses resulting from mold exposure is an incentive for anyone held liable for water damage to fix the problem. This is the argument used by Robert McGregor, a partner in a San Diego law firm, a specialist in litigation related to construction defects. The development of mold in a damaged building significantly increases the value of the case (both for the client and attorney), and by brandishing mold data—photographs, spore counts, and expert testimony—a skillful attorney can be successful in triggering insurance coverage. The strategy has been perfected by McGregor, who has recovered more than $50 million from contractors, building designers, and suppliers of construction materials. The practices of insurers are irrelevant to these cases, but by suing a building contractor, for example, any settlement is usually recovered from the company providing business insurance to the builder. (Ultimately, of course, the insurance industry passes on the cost to everyone by raising premiums.)

Attorney Mike Duffey explained his approach to mold cases at the mold litigation meeting in San Antonio. Perhaps a little carried away when making the point that scientific and medical studies have played little role in the majority of mold cases, he said, "I don't care if people are bleeding gallons an hour from their lungs." This, he argued, was irrelevant from a legal standpoint. Duffey explained that the key to winning a mold case was to demonstrate a pattern of "arrogance and indifference" on the part of the insurance company, builder, or other party that was being held responsible for the damage. This all made perfect sense, but I disagreed with him when he said of the conference audience that "Nobody is here because it's intellectually stimulating." He implied that the opportunity to learn how to secure $9 million in fees was the only sane reason for spending all day in a hotel room filled with attorneys. Although I'd concede that I would be willing to accept a check for $9 million, I was absolutely enthralled by the discussion. According to personality tests performed by employment consultants, many scientists and lawyers share similar profiles. Successful scientists and lawyers have a gift for advocating particular points of view, and I suppose that both professions attract individuals who feel compelled to wrestle with complicated questions. Despite this similarity, the best lawyers are far superior speakers to any biologist I've heard at a podium,[11] and I've listened to some of the famous ones including the late Steven Jay Gould, numerous Nobel laureates, Lynn Margulis, Craig Venter, and other genome stars (Venter wore the best suit). None of them come close to Mike Duffey or his partner Michael Childress. If you walk into a courtroom and find one of these individuals on the opposing side, settle.

Rich Finigan, president of the American Society of Professional Real Estate Inspectors, spotlights a number of modern building practices that promote mold growth.[12] Finished basements are very attractive to molds, because water seeping from the soil through the walls of the foundation can soak the paper-wrapped drywall yet remain invisible to the homeowner. Unfinished basements also get wet, but damp patches on walls are evident and there are fewer things like carpeting and furniture for the molds to feed on. The cement blocks of the foundation walls are supposed to be treated to limit water intrusion, but this is a common omission by builders. Ideally, a house should be built on ground that is raised above the surrounding yard so that water drains away from the foundation. But drive around any modern neighborhood and see how many homes have yards that slope from a high point behind the house toward the street. Vapor barriers made from thick plastic sheeting are another problem, because they block water that seeps into a wall cavity (the gap between the exterior and interior walls) from working its way into the living space where the leak is likely to be detected. Concealed from the occupants, the soaked wall cavity can fill with spores. The absence of a wall cavity causes even greater trouble. The builder of a mansion in Kentucky failed to finish the beautiful brick exterior so that it would repel water, and worse still, managed to push the interior wall up against the inner surface of the brick. Compelled by the laws of physics, the interior wall acted as a sponge, pulling rainwater through the porous brick to create an enormous Petri dish. This one made the evening news when it was burned to the ground.

Air-conditioning systems can oblige molds with a vacation spot. They love the pools formed by water condensing from the cold air. Similarly, forced-air heating systems drive warm air through the same ducts that carry chilled air in summer: Once clogged with cat hair and other household debris, the ducts trap fungi and treat them to a sauna. The nutrient content of the dust that collects in ducts must be quite low, so it is difficult to fathom how massive accumulations of mold spores develop in ducts. Most likely, they are formed on some cellulose-based building material and then sucked or blown into the ducts by the heating or air-conditioning system. (Cats are also apt to add to the nutrient content of ducts by a more direct mechanism.)

Some consultants blame mold growth on the demand for energy-efficient homes set off by the oil crisis in the 1970s, condemning hermetically sealed windows and wall cavity insulation for creating moisture traps. This seems like an excuse for poor workmanship: Insulation can be installed in such a way that any water that trickles into the wall cavity can drain toward the ground. If energy-efficient homes are to blame, one would have to argue that breezier dwellings would be dry and mold-free. Living in England for 24 years, where the construction of drafty homes was perfected by centuries of experimentation, I gained some personal knowledge of indoor wind patterns—as a student I lived in a house that lacked a back door—and attest that, despite their zephyrs, these dwellings were wet as sponges. (I'll discuss the mold situation in British homes in the next chapter.) Similarly, a drafty house isn't a cure for leaking pipes.[13]

Roof leaks are, of course, another cause for severe depression. As I have been writing this chapter, I have managed to associate every construction defect with some imperfection in my own house (just as I became convinced that I had a worm infestation when I wrote the immunology chapter). Pessimist that I am, I must concede that it's a very good day if one's worst fear (beyond the silence of death) is faulty roof shingling. (Sorry to spoil your day, Chicken Little, but, if you didn't know it already, there is a national problem with asphalt mineral surface shingles manufactured between 1986 and 1996.)

Unoccupied houses can be magnets for molds. The mere act of opening and closing doors and windows, or turning air-conditioning and heating on and off at appropriate times, and being aware of moldy smells offers excellent protection against *Stachybotrys* and its allies. I inspected a stone cottage in Ohio whose owner had closed all the windows, locked the front door, and left for a three-month vacation in France. While she had abandoned herself to the delights of Provençal cuisine, molds had munched their way from a basement leak throughout the entire house. Daytime temperature had peaked in the mid-90s for a couple of weeks that summer; with water vapor billowing from the basement, the cottage soon accommodated a mycological circus.

In addition to homes, *Stachybotrys* and other indoor molds include office buildings, schools, college residences, and warehouses in their diets. Experts know where to look for construction defects that are likely to lead to mold growth. These include poorly finished window surrounds and other faulty joints that are exposed to wind and rain. Keeping things very simple, we're talking about holes through which water can move into a building. A skilled builder will not make these kinds of mistakes, but the best carpentry cannot survive a leaking pipe or a hole in a roof. By documenting the building errors that resulted in an astonishing amount of mold growth in a high-rise apartment complex erected as a residence hall for students, attorney Robert McGregor won a $6.1 million judgment for San Diego State University. The entire building was shrink-wrapped in plastic while contaminated materials were removed, and then the structure was rebuilt in a manner that discouraged future mold growth. Although the value of "quality insurance" is questionable, there is no doubt that it's worth spending time finding a builder with an established reputation for "quality construction work."

What about "quality mold inspection"? Such a thing does exist. An industrial hygienist, or anyone else with experience with moldy homes, can locate areas of significant mold damage by poking around in basements, crawl spaces, closets, and other hidden places. Some inspection companies use a dog to do the sniffing. They have far superior noses and don't demand a benefit plan. The *Hamilton Journal News* blew the whistle—so to speak—on Hunter, a 2-year-old Border collie with expertise in mold detection, who works for Moss Restoration Services in Cincinnati.[14] Hunter can get into places where portly human inspectors can only dream of crawling, and will paw at a wall if he sniffs the perfume of a concealed mold. By swiftly locating sites where mold is growing, a mold dog can save a great deal of the collateral damage done by inspectors who may have to tear into lots of wall cavities before finding a noxious spot. But even when the contaminated areas are conspicuous, customers besotted by Hunter's television appearances demand a visit by the dog as part of the assessment process. Hunter is an adorable animal: When I interviewed Hunter's employer, Jim Moss, he—Hunter, not Jim—insisted on nuzzling me while I tried to maintain an air of distracted professionalism.

Bill Whitstine of the Florida Canine Academy was responsible for training Hunter. Though he charges $10,000 or more for each four-pawed mycologist, Whitstine believes the market will support at least 180 mold dogs. If he is wrong, the Academy will have invested a lot of time and dog food in producing pets that will be very bored by the daily routine of unemployment, and go berserk whenever their unwitting owners order a mushroom pizza. But he must be doing something right because he has trained bomb- and drug-sniffing dogs for the government, and claims to have shifted 90 termite dogs since 1997. A wonderful aspect of Whitstine's business is its reliance upon young dogs from animal shelters. He has found that a dog's temperament, rather than its pedigree, is a reliable indicator of its aptitude for detection work. Dogs sprung from death row are trained for 600 to 800 hours before graduation.

Commenting on the benefits of employing mold dogs, Whitstine says, "We don't have X-ray eyes so we can't see inside walls. But the dogs have X-ray noses if you will, and they can sniff inside the walls and tell us what's there."[15] The scent detected by a mold dog isn't known. Jim Moss says that Hunter can become confused by other fungi. The dog has embarrassed people suffering from athlete's foot by doing his "It's here! It's here!" performance next to their feet, making them wonder if they have stepped in something nasty. Apparently, Hunter's powers of detection are so strong that he can locate moldy toes inside socks and shoes. The mycological sensitivity of dogs is also put to use in truffle hunting, though pigs are more often celebrated for their truffling skills. Pigs may have a superior sense of smell, but they are difficult to control when they unearth a truffle. It has been suggested that truffles exude a volatile compound that resembles a sexual attractant secreted by female pigs: Dogs are immune to the charms of sows, so they are more easily coerced to surrender their finds to their masters. If you have any business acumen, I'm sure you've recognized the opportunity of a lifetime here. If pigs can be trained to hunt for truffles, they can surely be taught to hunt for indoor molds: They have an unmatched sense of smell, are happy with a diet of kitchen scraps, and because they won't get turned on by the perfume of *Stachybotrys*, they are likely to snuffle around contentedly without trying to mate with a moldy air-conditioner.

While we're on the topic of mold inspectors, I should mention that there is no nationally recognized training program that qualifies someone to be a mold inspector. Industrial hygienists attend a few classes on mold contamination and learn how to use an air sampler. A few of them will be taught how to recognize some of the common mold species with a microscope, but there are no degree courses in mold inspection. The best people in the field have years of experience and there are very few of them. For every hygienist with real knowledge of fungi, there is an abundant supply of well-meaning oafs, and the occasional mold bandit who will suggest you torch your home when a $5 shower curtain needs replacing. A significant problem comes from the current status of mycology in the United States and Europe. Allow me to explain.

Harry Stone, an environmental technologist based in Cincinnati, surprised me once by saying, "This is a great time to be a mycologist." I gave him an inquisitive look. "Really," he continued, "Can you think of any other time that was better for mycology?" With so many questions about mold and so much money involved, one might anticipate that American mycologists who had once been discontent with lives of quiet desperation—arriving at work each day in a 30-year-old Trans Am or hillbilly truck without a muffler— would have found themselves behind the wheels of silvery Cadillacs (or behind a chauffeur in one of those Rolls Royces with a built-in cigar humidor), speaking on cell phones to clients in Beverly Hills. (Don't laugh. Pet psychics have made millions.) Unfortunately, there have been few such exceptional lifestyle transformations in my field. The toxic mold crisis has certainly done a lot to bring fungi into the public consciousness, but this does not translate into a renaissance for the academic study of fungi. University courses dedicated to fungal biology have become rare, as positions for mycologists are subsumed by the demand for specialists in techniques, rather than experts on particular groups of organisms.[16] The majority of the well-funded labs in which fungi are studied obtain their support because they investigate diseases caused by fungi, or because the researchers use fungi as models for understanding fundamental cellular processes that are common to all eukaryotes (organisms, like us, whose cells contain nuclei). For example, one of the three recipients of the Nobel Prize for Medicine in 2001 was Sir Paul Nurse, a British researcher whose work on yeast has had far-reaching implications in understanding cell division and cancer. Though he works with a fungus, Sir Paul would never describe himself as a mycologist. Authorities on other groups of organisms such as nematode worms, insects, or particular families of plants have suffered the same dismissive attitude as mycologists in the academic world. Mycology remains a tough sell.

Personally, I have little reason to complain because I have been fortunate enough to have been employed as a mycologist since obtaining a doctorate. But to answer Harry's question, I think he's wrong. The golden age of mycology was over by the middle of the twentieth century. The mold problem has produced job openings in companies specializing in the remediation of mold problems, and opportunities for some expert witnessing for the rare mycologist able to function well in a courtroom. That such an esoteric subject as mycology has entered mainstream consciousness offers powerful evidence for the relevance of scientific literacy. But a deep knowledge of fungi is unnecessary when a bucket of bleach and a scrubbing brush may be the most important tools of the trade.

Most attorneys specializing in mold-related litigation possess little, if any, scientific training and are not equipped to assess the merits of the research on *Stachybotrys* and other indoor molds. They call on the testimony of experts most likely to support the claims made by their clients, and often attempt to sweep aside the profound uncertainties about the health impact of mold exposure discussed in this book. The courts remain very hesitant about mold-related illnesses, however, and little money has changed hands based on claimed illnesses caused by specific molds. There is considerable controversy about the relationship between mold exposure and pulmonary hemorrhage, and the case for an impact on mental functioning is immeasurably weaker. Remember that the viewpoint of the expert witnesses in the Ballard case concerning mold and brain damage was dismissed for lack of supporting scientific studies. When he rejected the expert testimony in the Ballard case, Judge Dietz referred to the U.S. Supreme Court's Daubert Opinion of 1993. Legal scholars view Daubert as one of the most important cases involving scientific evidence in the twentieth century. The Dauberts, and a second family, had sued Merrell Dow Pharmaceuticals, alleging that use of Dow's anti-nausea drug called Bendectin during pregnancy resulted in serious birth defects in their children. During the original trial in California, expert witnesses had discussed animal experiments that showed that the drug could damage a developing fetus, and presented statistical evidence gathered from mothers who had taken Bendectin, all of which supported the toxicity of the drug in humans. But Dow's attorneys argued that analysis of solid, peer-reviewed studies showed otherwise: Bendectin use was not a risk factor for birth defects. The California courts determined that the evidence against Bendectin did not reach the standard of "general acceptance" in the scientific community and sided with Dow. In reviewing this case, the Supreme Court threw out the standard of general acceptance, allowing testimony from experts who held viewpoints that differed from the prevailing opinion. If the judgment had ended there, the Justices would have given a voice to the countless lunatics that prowl the fringes of science, but they also ruled that trial judges must screen scientific evidence to ensure its relevance and reliability. In other words, some science is unreliable. The importance of the Daubert ruling is so far-reaching that I'll quote a few sentences:

> Cross-examination, presentation of contrary evidence, and careful instruction on the burden of proof, rather than wholesale exclusion under an uncompromising "general acceptance" standard, is the appropriate means by which evidence based on valid principles may be challenged. That even limited screening by the trial judge, on occasion, will prevent the jury from hearing of authentic scientific breakthroughs is simply a consequence of the fact that the Rules are not designed to seek cosmic understanding but, rather, to resolve legal disputes.[17]

Very clever, very pragmatic. Dietz used this argument to exclude the Ballards' personal injury lawsuit against Farmers Insurance for its role in Ronald Allison's "encephalopathy."

Where does this leave anyone who thinks they may have suffered brain damage caused by exposure to mycotoxins? There are a few reports in the scientific literature that describe neurological symptoms in people living in moldy environments, but these have not been subjected to the kind of rigorous case-controlled scrutiny applied to the pulmonary hemorrhage patients in Cleveland. Many of these are anecdotal studies of single patients. Here's a good example. After a 42-year-old radiologist suffered repeated bouts of bronchitis, he developed a hand tremor, twitching of arm muscles, and his vision deteriorated.[18] The putative source of the illness was identified when severe mold contamination was discovered in a room adjacent to his office. Once the mold damage was cleaned and the radiologist received antifungal medicine, his symptoms resolved. This story is riddled with problems. Though the timing of the symptoms and the discovery of the mold overlap, there is no evidence of a causal connection. That's why this is referred to as an anecdotal report. If mold toxins were responsible for the tremors, the prescription of an antifungal drug seems irrational: These pharmaceuticals are used to treat fungal *infections*, not as an antidote to fungal toxins. It's possible that the patient was poisoned by mycotoxins, but this study and others of its kind are clearly inadmissible in court.

Besides tremors, other supposed symptoms of mold exposure consistent with neurological damage include memory loss, seizures, hallucinations, and depression. Wayne Gordon (one of the expert witnesses involved in the Ballard case) and colleagues at Mount Sinai School of Medicine in New York City have been conducting careful studies on cognitive impairment in individuals who have reported problems after exposure to *Stachybotrys*. In one of their papers they discussed the results of detailed assessment of patients' verbal skills and memory.[19] The performance of the study group—who were highly educated middle-aged women—was compared with a large database of measurements from individuals who had not been screened in relation to mold exposure. (In other words, the researchers had no reason to think that those in the control group had been exposed to mold, though some of them might have been.) This is by no means a perfect experimental design, because the mental function of the women before their encounter with molds had not been measured, and there was no way of determining their level of exposure. The results indicated that all of the patients met the criteria for impairment based on a lower-than-anticipated score on at least one test. A variety of tests showed that 30 to 60 percent of the patients manifested measurable memory problems, difficulty concentrating, or showed some learning disability compared with controls. These are intriguing results, but more research is needed before anyone can say with certainty whether mycotoxins can cause mental impairment.

Some claims about the toxicity of mold seem reckless to anyone with even a passing interest in medicine, but this hasn't kept them from the courtrooms. *Linda Maxwell v. Pleasonton Unified School District* concerned mold and Tourette's syndrome.[20] A flood caused by a faucet that was left running all weekend resulted in mold growth in daughter Rheannon's school. Rheannon was diagnosed with Tourette's 10 days after spending time in one of the moldy classrooms. The jury was told that the girl was exposed to mold spores, and that this had caused her neurological illness. Tourette's syndrome is characterized by involuntary movements, including rapid blinking and facial ticks. It is usually diagnosed in young children and is a lifelong condition. At least some forms of Tourette's are thought to be genetic in origin, perhaps having nothing to do with environmental factors. At the trial, evidence was presented by the defense that Rheannon was showing symptoms on her first day back at school after the flood: If mold toxins were the cause, the effects must have been instantaneous. The attorneys representing the Maxwells couldn't find a neurologist willing to testify on the link between mold and Tourette's—no such relationship has ever appeared in the medical literature. The jury rejected the plaintiff's argument. So why did a parent make this appeal, and why did a law firm agree to support a patently frivolous claim? The initial $32 million judgment in the Ballard case made anything seem possible.

Some of you may be thinking that I'm being schizophrenic by siding with the researchers who think that *Stachybotrys* may have caused lung bleeding and by showing sympathy for some aspects of Melinda Ballard's crusade, while also agreeing that there has been a great deal of bad science and frivolous lawsuits. If so, you're wrong. What I have done is to argue consistently for logic over emotion, which is too often overlooked in the discussion of toxic mold. In a few years, careful scientific study is certain to lead us to the right conclusions. In the meantime, I think we should be very wary of dismissing the potential risk of mycotoxin exposure, and remember that troops of buffoons lie at both ends of the spectrum of moldy opinions.

Attorneys are waiting to see if indoor molds become the next breast implants or the next asbestos. Following a rash of multimillion-dollar lawsuits against the manufacturers of silicone breast implants in the 1990s, the unilateral redistribution of wealth was halted by continuing scientific research that refuted the original connection between the prosthetics and diverse illnesses. Asbestos is different. The courts are clogged with thousands of claims, some by people suffering from lung scarring or asbestosis, others from a very aggressive type of lung cancer called mesothelioma, still more by healthy people claiming to have been exposed to the mineral and fearing later disability. The value of these lawsuits may exceed an astonishing $200 billion, and the insurance industry will be asked to pay for about a third of that. There are plenty of questions about asbestos and lung damage, including uncertainty about safe levels of exposure (if there is any such thing) and the significance of the combination of fiber inhalation and cigarette smoking. But research on asbestos inhalation has gone on for decades, and peer-reviewed case-control studies demonstrating the risk associated with asbestos inhalation are seen as classics by epidemiologists. With so little definitive information on the hazards posed by mycotoxin-producing fungi, the future of mold lawsuits and the rest of the indoor mold industry is unclear, but the government is getting itchy.

The U.S. Chamber of Commerce sponsored a conference in 2003 attended by representatives of the building industry, the insurance industry, and a few law firms. Republican congressman Gary Miller of California was the keynote speaker and launched into a free-ranging narrative that criticized scientific research on indoor molds, accused Melinda Ballard of perjury, and ranted about the hardships faced by builders trying to construct homes on land occupied by endangered species.[21] The U.S. Chamber commissioned a pair of written studies on the mold problem.[22] One penned by attorneys Clifton Hutchinson and H. Robert Powell was titled "A New Plague—Mold Litigation: How Junk Science and Hysteria Built an Industry." They concluded, "Bad science and worse journalism are to blame for the mold fiasco."

Miller's evaluation of the mold crisis isn't endorsed by everyone on Capitol Hill. Representative John Conyers, Democrat of Michigan, has been concerned about the risk of *Stachybotrys* exposure for some time. In 2002 he introduced House Bill 5040, the "United States Toxic Mold Safety and Protection Act." House Bill 5040 is also known as the "Melina Bill," which refers to Melina Walker, the daughter of Conyers's office manager Pam. After *Stachybotrys* invaded the Walker home, Pam and her two daughters developed skin allergies and suffered from frequent nosebleeds. Seven-year-old Melina suffered severe asthma symptoms, losing much of her lung capacity by the time they fled the house. Pam Walker's insurance policy didn't cover her losses, and so she resorted to a lawsuit. This is such a familiar story: anecdotal evidence of illnesses caused by exposure to *Stachybotrys*, family forced to leave their home, failure of insurance company to compensate the homeowner, homeowner hires an attorney. The Melina Bill is designed to protect Americans from the physical and financial impact of toxic mold. It would generate federal funds to initiate "a comprehensive study of the health effects of indoor mold growth and toxic mold." When the research is completed, "not later than one year after the effective date of this Act," the EPA is expected to announce "national standards for mold inspection, mold remediation, testing the toxicity of mold, and protection of mold remediators." The bill calls for national certification of mold remediators and others in the mold business, and guidelines for identifying construction methods that, inadvertently, lead to mold growth. There's even a section on public education, so perhaps we may anticipate public service announcements by mycologists at sports events.

Returning to Earth, little of any scientific value can be achieved in a year, so the research part of Bill 5040 will have to be scrapped. Because nobody in the field can agree on standard methods for measuring mold or its toxins, or for graduating new mold remediators, I guess that will have to be discarded, too. But there's one part of the bill that will have an immediate effect. Everything I listed above falls under Title I, Research and Public Education. Title II covers Housing Provisions. If this were to pass, it would create mold millionaires and make investors in real estate consider emigration. The following procedures would become law: annual inspections of rental properties for mold damage, disclosure of existing mold growth before a home is offered for sale, and mold inspections before home sales or leases. (I haven't slept a wink since I read this section of the bill. How am I going to train all those mold pigs?) Other sections of the bill deal with standards for the construction industry, federal funds for correcting mold problems in government buildings, and tax breaks for those who incur the costs of mold work. But Title VI, the National Toxic Mold Hazard Insurance Program, is most pertinent to this chapter, because it will use taxpayers' money to cover the cost of mold damage claims. Insurance companies will no longer shoulder that burden. This means that we can all look forward to lower insurance rates—right?

State legislatures are also working on their own mold bills, seeking ways to protect insurers and consumers from each other and from *Stachybotrys* and bogus contractors. California Assembly Bill 284, authored by Hannah-Beth Jackson, Democrat from Ventura, required the California Research Bureau and Department of Health Services to study the health effects of indoor molds, the efficacy of the methods used to determine the extent of contamination, options for remediation, and other issues. Legislators demanding reliable and concrete answers are going to be disappointed. Mold is toxic. But which mold, and under what conditions, and when spore counts exceed how many per cubic meter? People like Pam Walker have no interest in waiting years for a definitive study that ferrets out the danger of indoor molds. She told a reporter from ABC News that she knows all she needs to about *Stachybotrys*, and will pass her key to anyone who wants to stay in her contaminated house in Detroit.[23]

The present contains nothing more than the past, and

what is found in the effect was already in the cause.

—Henri Bergson, *L'Évolution créatrice* (1907)

Everlasting Strife

Poor old *Stachybotrys*, it never meant any harm. Those black spores weren't made for flying around homes. Its spidery colonies evolved to munch outdoors. Other fungi—not humans—were the intended victims for its mycotoxins. One could pronounce similar innocence for the bacteria that cause food poisoning, black widow spiders, even grizzly bears. The fact that most of today's biology evolved without us in its cross hairs is of little comfort when a shark has your leg in its mouth, or when you're coughing up blood in a damp bedroom. There are a few exceptions to this prehuman picture of evolution: We do have some personalized enemies. These include infectious microbes that employ biochemical artifice to thwart our immune defenses or brush off prescription antibiotics, and far simpler self-replicating adversaries, including the mad-cow prion and human immunodeficiency virus, that offer the best hope for erasing us from the immediate future. Although I have acknowledged the mold's nastiness, it's clear that *Stachybotrys* would never have made mention alongside these menaces in the media's top 10 list of the bad bits of biology without skillful marketing.

To explain why there are molds that grow in our homes, I'll begin at the beginning, which was a very long time ago. In the beginning there was no life. (The Bible is unassailably correct on that point.) The Earth was formed around 4.5 billion years ago, and the first cells probably didn't put in an appearance for the first billion years of the planet's history. Whenever life was born from the primeval muck, we can be sure that it was born bacterial. Fast-forwarding a hundred quadrillion seconds from this purely micro-bial time, to the ecosystems of the twenty-first century, we find relatives of these aboriginal cells among today's bacteria. All con-temporary bacteria slot into two groups: the Eubacteria and the Archaebacteria (or Archaea). These organisms had the planet to themselves for 2½ billion years and evolved most of the fundamental biochemical mechanisms that made life possible for all of the fungi, plants, animals, and protists that would appear. Some bacteria met their energy needs by capturing carbon dioxide and releasing meth-ane, others powered themselves using compounds containing iron and sulfur, and brilliantly colored green and purple bacteria made food by photosynthesis. These self-sustaining bacterial ways of life persist today, along with a parallel bacterial world that feeds on things made by other organisms. Bacteria in this last group are called heterotrophs, and include the bacteria in our guts and those that cause disease. (Animals and fungi are heterotrophs, too.) Through this biochemical brilliance, bacteria have contrived to live every-where on the planet, from the coldest to the hottest, to the saltiest and most acidic and most alkaline places, to the stinkiest pits that humanity has fashioned—in short, to thrive in places as inhospi-table as the toilet in my stepson's apartment.[1] Even in this most frightful location, bacteria make a decent living above and below the malignant water. Beyond the bacteria, every other form of life constitutes icing on the toilet cake, nothing really substantial in bi-ochemical terms. Humans and molds are part of this sugary frosting, at least that's what the aliens will say when they come back and clean the globe down to the rocks.

Humans and fungi share a very ancient ancestor who was built from bacterial components. This was the first eukaryotic cell, a modified bacterium whose genes were protected in an envelope called a nucleus. At some point in the evolution of eukaryotes, an-other bacterium became housed inside the first one, was trans-formed into a powerplant called a mitochondrion, and served the energy needs of the hybrid cell. Bacterial cell inside another bacterial cell: Think Russian doll, and you have captured the soul of most eukaryotes. Nobody is sure when this innovative cell type appeared, but fossils of eukaryote cells have been dated to about a billion or so years ago, when 70 percent of the Earth's history—from today's perspective—was already over. Fungi, plants, animals, and protists are eukaryotes. When we view the pageantry of life on this gigantic scale, where we fit everything into a handful of kingdoms,[2] we discover that fungi and animals are surprisingly close relatives. Puffballs and puff adders work in essentially the same fashion and sprang from a common ancestor. This ancestor of the fungi and animals resembled a microscopic choanoflagellate (*co-an-o-fladge-l-ate*), an aquatic protist with a tail or flagellum mounted at the base of a see-through collar. It's easy to picture one of these cells with a violent thought experiment. Begin by thinking about one of those surgical collars taped to a dog to prevent it from pulling out stitches; then remove the dog's head, and stuff one end of a garden hose in the neck stump, and finally, replace the rest of the dog with a blob. Now, grab the free end of the garden hose and wiggle it around. Apart from the difference in size, you have created a perfect model of a choanoflagellate. These simple organisms feed by lashing the garden hose/flagellum, creating a vortex that drags particles down into the base of the collar where tasty morsels are filtered out for consumption. So how do we get from there to a black mold, and to a biologist sitting in a writing shed? Elementary, my dear Watson: hundreds of millions of years plus natural selection. Cells that look very similar to choanoflagellates form part of the structure of sponges, and sponges are some of the simplest multicellular animals. The simplest fungi are called chytrids (*ki-trids*). Chytrids form single cells with flagella called zoospores, and many mycologists equate these with the ancestral choanoflagellate. Biologists in writing sheds and other animals developed along a different line from the chytrids, but assuredly sprang from some unicelled neighbor of the choanoflagellates. There is certainly speculation involved in this story, but the overall picture is supported both by the structure of contemporary organisms, and by their genes.

Humans evolved much later than molds, of course. The fossil record reveals that recognizable fungi were flourishing around 400 million years ago, the same time that plants and insects show up on land; *Homo sapiens* didn't arrive for another 399.8 million years. *Stachybotrys* fossils—or spores that look like those of *Stachybotrys*—have not been found, but other black molds are beautifully preserved in amber. Amber is a fossilized form of tree resin, and anything trapped in resin is preserved along with the goo. Melanized spores and hyphae of black molds are found in amber that was preserved tens of millions of years ago. The most beautifully preserved of these fungi are species recognized as sooty molds, which have been described by a group of Finnish and German scientists.[3] The fossils are found in European amber that formed between 22 and 55 million years ago. I want to tell you about sooty mold fossils, because they may hold some clues to the behavior of some of the outdoor molds before they had an opportunity to move in with us.

Sooty molds grow in abundance on the surface of leaves and stems, sometimes blackening an entire tree. On leaves, the molds form sticky webs of hyphae; on the trunks of trees, they can accumulate in the form of thick, lumpy cushions. Unlike pathogens that penetrate plant tissues, sooty molds content themselves by bathing on the outside of the host. Their food comes from two sources. The leaves of many plant species are tacky because they leak sugar-rich sap. This allows sooty molds to grow on walnut, hibiscus, and catalpa (the stately tree with huge leaves and dangling pods). Insects are also vital to the development of sooty molds. Scale insects and aphids suck the juices from plants and then express droplets of the sap from their rear ends. This exudate is called honeydew, and transforms the leaf surface from a dry desert into a lush garden for microbes. Abundant sooty mold growth can harm plants by preventing photosynthesis by shading the green leaf cells with their black hyphae. The same type of fungi show up frequently on patio umbrellas and other pieces of outdoor furniture sweetened by tree sap. The term "sooty mold" can be applied loosely to any fungus that creates a black powdery-looking growth, but taxonomists usually reserve this for species from a particular group of fungi called the Capnodiales, which have a very distinctive appearance under a microscope. Unlike the straight-sided hyphae of most fungi, these species have tapered filaments that look like strings of black pearls. The amber fossils appear identical to living species with this structure, demonstrating that modern looking sooty molds have been around for millions of years. The researchers who described the mold fossils also determined the source of the amber. It came from a species of pine tree that is now extinct. Resin slowly drizzling down the trunk of the pine tree drowned the molds clinging to its bark.

Though *Stachybotrys* and other indoor fungi blacken surfaces, their cells don't look like the pearly filaments of a sooty mold. But indoor fungi and sooty molds on plants do share two significant features: both have jet-black cell walls, and both are virtuosos in the arts of water-retention and surviving dehydration. I'll discuss coloration first.

Why is black pigmentation such an important feature of sooty molds and other black molds? Without blackness, a sooty mold would be a translucent mold, and it would not survive for long on sunlit leaves. Like animals, fungi use melanin as a sunscreen, particularly as a protection against the damaging ultraviolet wavelengths of light. Fungal melanins (there are different kinds) are enormous molecules built from ring upon ring of carbon atoms that interconnect to form a highly absorbent chemical maze. The resulting blackness of the cell is caused by the ability of the pigment to absorb all of the visible wavelengths of sunlight, but what isn't so obvious is the fact that the melanin is also effective at sopping up the damaging energy conveyed from the sun by gamma rays and X-rays. Melanin absorbs the energy from these forms of radiation as they pass into the black cell wall and channels it deep into its dense molecular structure. Apologies to chemists for simplifying things a bit here, but some of the energy is dissipated by molecular vibrations and other chemical adjustments, and what's left of the influx is released as heat. This gentle warming of a cell is far preferable to the chromosomal havoc caused by ultraviolet light and other slices of the electromagnetic spectrum. Survival without melanin, or some other pigment molecule, is inconceivable for an organism living on land. British landscape painter J. M. W. Turner (1775–1851) once said, "If I could find anything blacker than black, I'd use it." No organism has ever found anything blacker than melanin. (As a biologist, I have no difficulty understanding why black skin is superior to the kind of chalky drapery that conceals my innards.)

Blocking sunlight isn't the only thing that melanin does for a fungus. I spent a good part of the 1990s studying the effects of melanin production in fungi that infect plants, and in those that infect humans. The most frustrating part of the melanin research was that almost every laboratory that decided to look at the pigment in a black fungus soon "discovered" that the chemical had a new function: Melanin does this; melanin does that; but wait a moment, this is what it really does. When fungi are robbed of their black pigmentation, either through mutation or by deliberate poisoning of their metabolism, they don't perform very effectively. So if you look at the effects of melanin on the speed at which fungal hyphae grow through tissues, you'll find that the black ones penetrate faster, and when you look at the ability of the same strains to survive temperature extremes, sure enough they are hardier. Without melanin, the albinos grow slowly, cause less tissue damage if they are pathogens, are killed by lower doses of radiation and environmental toxins, and so on. The traits associated with skin pigmentation in humans offer a useful analogy. In many major cities in the United States, significant disparities in economic prosperity correlate with skin color. Though the depth of one's pigmentation is accompanied by other subtle genetic differences, the relationship between color and affluence (and all of the attendant consequences of wealth or privation) has nothing directly to do—physiologically speaking—with the production of melanin. Similarly, the list of melanin's attributes among fungi include many characteristics that are remote rather than direct consequences of being black.

It is difficult to pick out the few direct effects of melanization on fungi besides the resistance it confers to ultraviolet light. One characteristic that interests me is the change in permeability caused by the deposition of melanin in the cell wall.[4] Melanin may help a cell keep hold of its precious bodily fluids (in the words of General Buck Turgidson in *Dr. Strangelove*) and exclude the corrupting influences of the outside world. Speaking scientifically, this means keeping molecules dissolved in the cytoplasm inside the cell and keeping toxins out. Now it's time for some lateral thinking. Could melanin protect a cell from itself? *Stachybotrys* produces a slurry of mycotoxins in the watery interior of the spore and then releases them onto the cell surface. If the cell doing this secretion is a developing spore, then the mycotoxins end up in the wall of the spore. Because spores form in air, the toxins are likely to stay on the outside of the cell as long as the spore is kept aloft. But as soon as the spore hits a fluid environment—on a moist leaf surface, wet soil, paper, skin, or lung linings—the toxins will be dispersed, and some could reenter the cytoplasm. This could be very bad news for a mold. The toxins identified in *Stachybotrys* affect fundamental cellular processes such as protein synthesis, and it is likely that the mold is sensitive to its own poisons. While the mycotoxins are developing in the cytoplasm, they are likely to be sequestered in some way, held in membrane-bound vesicles, for example, so that they do not come in contact with the enzymatic machinery that keeps the toxin-producing cell alive. By depositing a layer of melanin in its wall, the fungus traps the toxins on the outside so that they can't worm their way back in. There is an obvious test for this idea. If I'm correct, a toxic strain of *Stachybotrys* in which melanin is inhibited should poison itself.

Given the likelihood that its blackness is key to the success of *Stachybotrys*, could chemicals that block melanin production prove an effective remedy for mold damage? The same idea has been floated by scientists treating human infections by black molds.[5] There might be a few situations in which indoor mold contamination could be addressed by fungicide application, but as long as bleach is effective at killing fungi, and mending plumbing leaks stops them from returning, higher tech treatments for mold growth face some stiff competition. A profitable market might be found by looking at products that are designed to get wet. Plastics and textiles impregnated with antifungal compounds could be used to manufacture mold-resistant shower curtains, tiling grout, outdoor furniture, or any other normally mold-prone item.

Returning to the fossilized fungi, I said that today's indoor molds shared a second characteristic with these prehistoric microbes. Evidence of ancient sooty molds demonstrates that fungi learned how to adapt to the desert-like surfaces of terrestrial plants a very long time ago. By evolving to cope with the wretched environment offered by the exterior of a plant leaf—moist for a couple of hours in the morning, followed by a day of withering dryness (we've all known someone with these characteristics)—other kinds of black molds, including the ancestors of *Stachybotrys*, were blindly prepared for attacking buildings. All but the wettest homes have dry spells; times when flood water recedes from the basement and rooms are aired. Drying of a wall means death to a fungus unable to deal with dehydration, and *Stachybotrys* is particularly good at growing when water is available and sitting tight when the water disappears. So indoor molds didn't evolve as indoor molds but just happened to have spent millions of years developing a suite of natural characteristics that also allowed them to exploit water-damaged building materials when we turned trees into houses. Mycologists encounter similar examples of fungal opportunism all the time. Consider a species of mold called *Geotrichum* that eats CDs, particularly in tropical countries, pitting the plastic and metal parts of the disc, leaving them unplayable.[6] The 'same fungus also grows on a variety of synthetic materials, but its ability to degrade these products wasn't tested explicitly during the evolution of its biodegrading talents. *Geotrichum* thrives in the modern human environment because it possessed an unusually high degree of nutritional flexibility before we came along. This also explains why raccoons and opossums live in our cities. It is possible that natural selection has enabled some fungi to elaborate specific mechanisms to destroy our inventions, but I don't think it's necessary to invoke this. Instead, the ubiquity of fungal spores, and tremendous diversity of species and strains floating in the air, has allowed these microbes to taste-test everything in their environment. Because the chemistry of the substances they encounter determines whether they perish or grow, they have found lots of novel food sources since the industrial revolution.

A diet of cellulose with very little else is a major challenge for a mold that grows on damp wallpaper. *Stachybotrys* is one of the few fungi able to cope with such meager gruel. Most fungi want something else: nitrogen. *Stachybotrys* needs nitrogen, too, but it is thought that it can thrive with very little of the stuff compared with the indoor species of *Aspergillus*, *Cladosporium*, and *Penicillium*. For this reason, when *Stachybotrys* finds its way onto wet, paper-clad wallboard, soggy clothing, or other objects manufactured from plant fibers, it sometimes has an open field. Once its spores germinate and a mycelium develops, its mycotoxins are thought to ward off any competing molds and bacteria,[7] and so its hyphae chew warrens through the surface layers of a wall and spawn billions of toxic spores. In badly water-damaged homes, *Stachybotrys* can establish itself over a large area such that decomposing walls become spattered with its conidia and little else. In other situations, *Stachybotrys* is one among many contaminants, and mycologists have found that a succession of different fungi colonizes the building materials. When this happens, *Stachybotrys* is the last in line, unable to compete with faster-growing species until they can no longer sustain themselves. According to this scenario, the most toxic of the indoor molds is comparable to the vultures that pick the last fragments of meat from an antelope carcass after the lions and jackals have left the scene.[8] Its extreme toxicity may be viewed as an insurance policy against any other scavengers who might otherwise scrape a living in territory staked by *Stachybotrys*.

Beyond these physiological adaptations that allow *Stachybotrys* to grow where few others can, and the possibility discussed in the previous chapter that modern building practices exacerbate mold problems, there may be other reasons for its success. In 2002, I visited fellow mycologist Ian Ross. Ian is a fortunate man. Born in Britain, he escaped to the United States in the 1960s and became a professor at the University of California in Santa Barbara. The decision to pursue an academic career on the West Coast—rather than a life of probable desperation on a damp concrete campus on the sceptered isle—wasn't difficult. His commute offered a view of the sparkling Pacific, along with the frequent distraction posed by bikinied coeds walking to the campus surfing beach. After more than 30 years teaching mycology and supervising students in his lab, he retired. But his mycological career wasn't over. Within weeks he was hired as an expert witness for a mold-damage case. Having honed his act in front of thousands of undergraduates, a jury substituted for another freshman class who needed to learn something about fungi (whether they liked it or not). Returning to the reason for my visit to Santa Barbara, I wanted to hear what Ian thought about *Stachybotrys*. After all, having thought about fungi for all of his professional life, I figured that he must have a few things to say about this toxic one in particular—especially since he'd been planning his lecture for the mold jurors. At one point in our discussions, he said that *Stachybotrys* might be spread by insects. Fungi with sticky spores are often dispersed by insects; *Stachybotrys* has sticky spores. When I asked Ian what mycological luminary had made this decree, he left the living room for a minute, then returned with a copy of a book he published in 1979. He read:

> Nearly all sticky-spored fungi may be accidentally dispersed by insects as the insects forage in the vicinity and brush against the spore mass, but many such forms have insect attractors in the fluid that cause the insect to eat the spore mass.[9]

The significance of insects in carrying sticky spores has been recognized by mycologists for a long time. The fungi that cause Dutch elm disease (species of *Ophiostoma*) are dispersed by insects. Their spores are produced inside the tunnels of wood-boring *Scolytus* beetles and are elevated on bundled stalks that they brush against the bodies of the insects. Could a similar mechanism apply to *Stachybotrys*? We explored the idea further with the aid of some wonderful local wine.[10] Building consultants are often surprised by the extent of mold growth inside wall cavities. It is possible that spores of *Stachybotrys* enter these spaces during construction, but this is questionable because our microscopic Darth Vader isn't a major constituent of the spores floating around in the air. So, we wondered, might common insect pests like pharaoh ants track spores into wall cavities as they make their way between the outdoor and indoor environment? Viewed with a microscope, the surface of a colony of *Stachybotrys* looks like a forest of stalks, each bearing a glistening droplet of fluid. The droplets develop in humid air when water condenses on mucilage associated with the developing spores, so each globule bathes a cluster of spores (plate 6). The stalks are very short, but perfectly placed to transfer spore-globs onto the limbs of a moving insect. *Stachybotrys* is different from most other indoor molds. *Aspergillus*, for example, forms bone-dry spores that seem much better adapted to dispersal by air currents. The possibility that *Stachybotrys* collaborates with insects—passively, or actively with the involvement of a chemical lure—adds a new dimension to the biology of this fungus. Could we deflect this particularly toxic mold by keeping indoor ants outdoors? Probably not, but Ian's idea is certainly worth a few experiments.

If *Stachybotrys* and other molds that grow in homes existed long before there were any buildings, it is reasonable to wonder why they didn't justify a story in any newspaper a few years ago. The answer preferred by the insurance industry, some researchers, and the CDC is that molds had always flourished in homes but industrial hygienists didn't look for them until recently.[11] Interest in indoor molds has a slightly longer history in Europe than in the United States, but even in Scandinavia, where indoor air quality is an obsession, public health scientists ignored molds before the 1980s.[12] The implication of the idea that indoor molds have been around forever is that the present excitement about *Stachybotrys* is misguided. Fatigued by the media attention given to the mold problem, many people express little sympathy for those who may have been sickened by it. I'll allow a skeptical physician in Cincinnati to speak for this constituency. When I discussed my interest in the Cleveland cases with this gentleman, he shook his head. "Why is this generation [he didn't specify whether he meant the babies or their parents] so screwed-up? Nobody complained about mold when we were kids." Mr. Bounderby, the banker in Dickens' *Hard Times*, would certainly have concurred. He recalled that his early childhood had been quite distressing: "I passed the day in a ditch, and the night in a pigsty. . . . Not that a ditch was new to me, for I was born in a ditch." Somehow, by sheer force of will, Bounderby was able to "pull through it," enabling him to become a wealthy man who held scant interest in the sufferings of anyone else. Like the physician, Bounderby would have viewed lung bleeding as a sign of self-indulgence. Media coverage of frivolous mold-related lawsuits has irritated many of those who express versions of the same sentiment, but it is unfortunate that IPH babies and their families are classed alongside people attempting to use *Stachybotrys* to finance an early retirement. Either the recent instances of lung bleeding in babies living in water-damaged homes mark a new public health problem, or the Cleveland investigation alerted physicians to an uncommon illness that has afflicted infants for centuries.

Interest in *Stachybotrys* will evaporate if innovative research leads us to the unshakeable conclusion that the lung-bleeding cases in Cleveland were caused by something other than mycotoxins. But what if there is another twist to the home contamination story? What if this particularly toxic mold is a recent North American immigrant? Some mold researchers speculate that *Stachybotrys* began moving across the continent in the 1990s, and by moving from moldy home to moldy home has probably spread to every state and province. In an article in the journal *McIlvania*, published by the North American Mycological Association, Bill Freedman showed a map of cases of putative mycotoxin-related illnesses in the United States and Canada. He concluded that "the mold is essentially found in buildings along bodies of water."[13] My own painstaking analysis shows that many Americans are likewise "found in buildings along bodies of water," so I don't think that Freedman can claim any great breakthrough in mold science. Dorr Dearborn, the physician who treated the pulmonary hemorrhage cases in Cleveland, is interested in the possibility that toxic strains of *Stachybotrys* may have been imported from Canada. Cleveland is a major port, and it is possible that the fungus arrived on moldy agricultural products, or on any number of different kinds of packaging material. Steve Vesper at the EPA takes this a step farther by suggesting that the mold began its trip westward in the middle of the twentieth century from Ukraine and eastern European countries where it had caused animal and human stachybotryotoxicosis. Given that the stickiness of its spores argues against habitual dispersal by air, perhaps it moved into North America on imported animals or animal parts, food products, or insects.[14]

There have been no lawsuits triggered by *Stachybotrys* in the United Kingdom. This isn't explained by the fact that the British are less litigious, at least not completely. Despite the availability of millions of very damp homes,[15] this particular mold doesn't flourish there. Other indoor fungi grow in abundance, but not *Stachybotrys*. There are differences in building practices, and in my humble view, houses in Britain are now being built with greater care than new homes in the United States. More expensive American homes are certainly bigger than cheaper ones, but the quality of the craftsmanship is usually indistinguishable. Why, for example, is it so rare to find a rigid wooden banister on a staircase in this country? Test this for yourself the next time you're invited to cocktails at the local car dealer's chateau: Amazing indoor pool, but I bet he has a wobbly newel. Most of the mold problems addressed in recent lawsuits have involved newly constructed homes,[16] but experts agree that the worst cases of mold damage are usually seen in older homes with severe water damage. In any case, I'm not convinced that the quality of homes old or new is the explanation for this geographical divergence. Even if they are better built, British homes are awfully humid.[17] Differences in the types of building material may be significant. *Stachybotrys* adores the type of paper-wrapped gypsum board used in American homes. Once soaked, gypsum board will leach calcium into the paper: The resulting combination of moisture, calcium, and cellulose provides ideal growth conditions for *Stachybotrys*. Temperature is another consideration. *Stachybotrys* likes being wet and warm: Remember that the first rash of lung bleeding cases in Cleveland was probably provoked by massive mold growth after an unusually wet, but normally warm summer. Temperatures in Britain rarely hit the 90s and never remain that high for days on end. Otherwise, given the humidity, London's biggest claim to fame would be its Hanging Mold Gardens. And there's another factor that *Stachybotrys* finds discouraging. Northern Europeans put on more clothing indoors when the weather turns cold; Americans adjust the thermostat. The outbreaks of stachybotryotoxicosis in Ukraine and in eastern Europe might be explained by the warming effect of other microbes in straw.[18] Cases of home contamination in these areas have not been described in the literature.

Now that we are becoming accustomed to regard climate as a malleable feature of our environment, it's worth considering whether the mold epidemic will wax or wane over the next decades.[19] Heavy rainfall in California in recent years, related to climatic swings and roundabouts caused by Pacific warming (El Niño) and cooling (La Niña), has already been implicated in mold problems, and others have speculated that the global trend toward higher temperatures may be playing a role in the spread of indoor molds.[20] There isn't any solid science behind these fears, but I suppose that some regions may experience worsening mold problems while others will be blessed by the future disappearance of *Stachybotrys*. The strongest evidence linking changes in fungal populations to global warming comes from the Netherlands. Analysis of data on lichen distribution in the Utrecht region since the 1970s is consistent with an increase in species that show a preference for warmer conditions.[21] If global temperatures continue to climb, however, we'll be dealing with more consequential challenges than lichen migration or indoor molds.

Recognition that indoor molds are part of the natural world doesn't affect the wisdom of avoiding significant exposure to *Stachybotrys*. After all, grizzly bears are natural, organically grown animals, but I don't want to live with them. But as I have explained in previous chapters, peoples' distress is often exaggerated. A phenomenon called social amplification plays an important role in fostering concern. Once clothed, housed, well fed, and in no apparent danger of being axed to death by a neighbor, human beings fixate on the silliest things. Let's use me as an example. I am, of course, quite likely to be killed in an automobile crash but never give this real threat any thought as I floor it down the freeway talking on my cell phone, drinking coffee, eating a burrito, cleaning my thick-lensed glasses, and so on. The flight I have to catch troubles me a little, 2 points on a 10 point scale, but everything is blotted out that evening when I look in the mirror in the hotel bathroom. As a possible signature of disorderly cells, the appearance of a red fleck on my forehead counts as a 6 or 7, because this could mark the fast descent toward diagnosis of something terminal. (It turned out to be hot sauce from the burrito.) Homeowner psychology has a major influence on the way that people perceive mold problems and whether they complain of breathing problems and other illnesses in contaminated homes. Analysis of questionnaires given to residents of water-damaged buildings reveals that individuals coping with the aftermath of a divorce or other traumatic events are especially prone to feeling anxious about fungal growth in their homes and tend to report a greater range and intensity of symptoms they ascribe to mold exposure.[22] This finding doesn't say anything about the actual hazard of mycotoxin exposure, of course, only that psychosomatic factors influence the symptoms reported by residents of contaminated homes. Careful analysis of the symptoms reported by participants in mold studies also shows reporting biases that are due to behavior rather than psychology.[23] For instance, parents who are smokers are more likely to underreport how often their children cough than parents who are nonsmokers. Similarly, irrespective of the degree of mold contamination in a home, smokers are less likely to report visible fungal damage, especially in comparison with people who suffer from allergies. These types of distortion make it very difficult to assess studies linking indoor molds and different illnesses that have been based on questionnaires.

Toxicologist Janet Weiss did a fascinating comparison between real and imagined risks by looking at public and scientific perceptions of *Stachybotrys chartarum* and *Staphylococcus aureus*.[24] *Staphylococcus aureus* is a bacterium that causes food poisoning, toxic shock syndrome, blistering of large areas of skin, and a plethora of infections. It is found in all places occupied by humans, affects the lives of millions of Americans every year (mostly through food poisoning), and our encounters with the bacterium can be lethal. All scientists and physicians agree that it is, in short, a horrible bastard. But other than the experts, very few people are at all concerned about *Staphylococcus*, and the bacterium holds close to zero interest for lawyers or the media. The comparison with *Stachybotrys* is astonishing. Despite the fact that the toxicity of the fungus may affect human health, nobody is claiming that it has killed more than a handful of patients in this country. Yet almost everyone has heard of the killer mold, and it has proven a boon for the legal profession.

I doubt that the survey of the natural history of molds and the battles between scientists, attorneys, homeowners, insurance companies, and politicians I've offered in this book will help us deal with *Stachybotrys* or with one another in a more sensible fashion. But I do think I have presented a reasonably objective viewpoint about the contemporary mold crisis. More importantly, I have argued that science needs some time to differentiate between the real and perceived danger of mold exposure.

. . . dry rot and wet rot and all the silent rots

that rot in neglected roof and cellar . . .

addressed themselves faintly to my sense of smell. . . .

—Dickens, *Great Expectations* (1860–1861)

A Plague upon Your House

In the last decade, investigative reports on television have been successful in worrying millions of Americans about a variety of domestic hazards including aluminum and Teflon in cookware, traces of pesticides on apples, radon, carbon monoxide, and microwave ovens. (I cannot say "needlessly worrying millions of Americans" with any confidence, because I now do all of my cooking in earthenware pots.) Some scientists view *Stachybotrys* as nothing more than an addition to the list of also-rans among this field of monsters assembled by media attention. I don't share this viewpoint, and have said all I want to about mine. This book closes with the news that *Stachybotrys* isn't the only fungus attacking our property. In the last chapter I took you to Santa Barbara, California, on the trail of the blackest of molds, and we return to the city now, because it is also cursed by dry rot.

Santa Barbara is such a stunningly beautiful place that it seems reasonable that this apparent utopia is troubled by one or two things. The realization that Michael Jackson's Neverland is just down the road is disquieting, but the window bars and security fences protecting houses show that their proximity to this strange estate isn't the only thing making Santa Barbarians paranoid. Daily news reports highlighting the frequency of home break-ins and violent crimes—with special emphasis on misdeeds committed by illegal immigrants—provide certain inspiration for purchasers of home security systems. But though alarms, drawbridges, and portcullises may rebuff criminals, they are no barrier to the dry rot fungus (plate 7). This pestilence can transform almost any home into rubble and sawdust.

I'll begin with the thoughts of journalist Matt Kettmann, who wrote a story for Santa Barbara's newspaper, *The Independent*, titled "Invasion of the House-Eating Fungus."[1] Matt interviewed the Kastner family who had moved into a $1.2 million ranch-style house. The nastiness began when "a mysterious foam slowly started peeking out of the laundry room walls." Dry rot was diagnosed and treated by an extermination company. Feeling confident that the fungus was gone, the family left for a vacation. When they returned, Matt explained, nearly a quarter of the house "was eaten in a mere week." Christina Kastner remarked, "It came back with a vengeance—like it got mad." The problem was solved once the exterminators found a "taproot with the diameter of a grapefruit in the rear of the house." This umbilical chord was cut, a ditch was dug around the foundations and filled with concrete, and the damaged home was refurbished. The Kastners' insurance policy didn't cover any of the $100,000 repair bill.

Many species of fungi cause the type of wood decay encountered in homes, but most of them prefer to work outdoors in woodlands. Very few specialize in indoor work. Much of the horrific damage reported in California is caused by *Meruliporia (marr-ul-ee-pore-ear) incrassata*. A different species, called *Serpula lacrymans*, is responsible for the same kind of destruction in Britain, elsewhere in Europe, and in other parts of the world, including Australia and Japan. *Serpula* means serpent or worm, evoking the taproots of the fungus that slither indoors. The specific name, *lacrymans*, is the Latin term for weeping and refers to the appearance of globules of fluid on the surface of the fungus.[2] Until a few years ago, the fungus was called *Merulius lacrymans*. (How do taxonomists ever settle upon names for their children or pets?) *Merulius* referred to the yellow color of the mycelium, which was comparable to the beak of a male blackbird (*Turdus merula*). I mention these etymological details to illustrate the vivid pictures that sometimes hide in the scientific names of organisms: glistening, yellowish colonies with serpentine taproots.[3]

The warmest invitation for dry rot is a house in which some wooden structure makes direct contact with the soil. A post sunk into the ground without the protection of concrete is a choice entry point. Dry rot also intrudes when soil accumulates under the bottom edge of a stucco exterior, or contacts the exterior particle board underneath vinyl siding. Many cases of dry rot could be avoided if builders ensured that a few inches of bare concrete showed all the way around a home exterior. (This is a good pointer for inspecting any home before purchase.) Even in regions without dry rot, piling soil against a house is an inexcusable mistake, because ants and termites can use the same pathways into homes as fungi. Other defects are less easily spotted. Cracks in concrete slabs and any holes for pipes or electrical work also serve as gateways for the fungus. Terry Amburgey, Professor of Forest Products at Mississippi State University, says that the dry rot fungus "will infiltrate a foundation, wood, or concrete, and pretty soon the entire house goes."[4] Comparing dry rot to "a horde of army ants in search of food," mycologist David Arora says that the fungus will "overrun everything in [its] way: bricks, stones, tiles, plaster, drainpipes, wires, leather boots, cement floors, books, tea kettles, even corpses."[5]

Indoor molds like *Stachybotrys* get their water from the dampness of their surroundings, accounting for their appearance in water-damaged homes. Dry rot operates in a completely different fashion. Its common name refers to the ability of the fungus to destroy wood in dry buildings, but ironically, *Meruliporia* and *Serpula* are very sensitive to dehydration. They thrive in the wet interior of a beam, but only emerge on surfaces if they remain damp and shaded from sunlight (which partly explains some of their fondness for basements). The key to understanding the extraordinary destructive abilities of these organisms lies in the taproot structure mentioned above. *Meruliporia*, *Serpula*, and other wood-decay fungi form two types of root-like structures, called strands and rhizomorphs.[6] Strands show little internal differentiation, though a thick "parent" hypha occupies the core of the strand and is surrounded by thinner hyphae growing along it in the same direction. To further consolidate the strand, the parent produces tendrils that wind around the other hyphae. Many fungi form these cables, even in culture plates. Experiments suggest that strands begin to develop in

A Plague upon Your House

129

response to nutrient exhaustion. According to this idea, strands are exploratory structures that the fungus can send out from the colony in search of new food sources. As soon as the scouting strand hits some nutrients, hyphae fan out to form new mycelia and begin extracting nutrients. It seems that as long as food is limited, the hyphae will cooperate with one another to grow across barren terrain, but as soon as they reach an oasis the strand disassembles. Rhizomorphs are a more complex form of strand: They are much larger, have a waterproofed surface, and enclose a central pipe for the transmission of water and dissolved nutrients. These are the structures that invade Californian homes (plate 8). The rhizomorph can grow for many meters, piping water from its base (in wet soil) to its tip (buried in dry wood or a slab of masonry). In some situations, air can pass along the rhizomorph so that the fungus can avoid suffocation while exploring the anoxic interior of a painted beam. Rhizomorphs bear some similarities to mushrooms: Both are complex, multicelled structures and develop through the interweaving and adhesion of scores of hyphae. Interestingly, a cut rhizomorph exudes a distinctive mushroomy smell. (Or, I suppose, a cut mushroom has a distinctive rhizomorphy smell.) The rest of the fungus has a similarly strong odor, which has led some home inspectors in Scandinavia to use dogs to detect dry rot before too much damage has occurred. (This predates the use of mold dogs for the detection of indoor molds, which was discussed in chapter 6.) A recent study in Finland evaluated the skills of Labrador retrievers.[7] The investigators concealed small pieces of pine colonized by *Serpula* and other wood-rotting fungi, plus cultures of indoor molds and bacteria. Clean pieces of wood were hidden as controls. The dogs were very effective at finding the contaminated materials and cultures, but failed to discriminate between the various microbes—which is probable cause for celebration among human mycologists. Though difficult to detect during the early stages of installation, advanced cases of dry rot are unmistakable: Nothing else produces white mats of mycelium clinging to the surfaces of beams, strands of hyphae and rhizomorphs dangling between different areas of decay, and revolting fruiting bodies.

Once in a basement, hyphae of dry rot fungi pulse through the microscopic cells of hardwood beams, digesting their cellulose walls, and transforming springy lumber into parched brown cubes that crumble when touched. Some contractors refer to dry rot as "brown cubical rot." Brown rot fungi in forests leave the same signature in decaying logs. The brown color is due to the progressive concentration of a dark-pigmented material called lignin, which the fungus leaves behind as it digests the cellulose. (White rot fungi extract the lignin and other pigmented components of the wood, leaving white cellulose behind.) Unlike indoor molds that rely upon spores to move from one location to another, dry rot fungi can colonize an entire home without forming a single spore. If it can conduct enough water indoors through its rhizomorph, or if the home is saturated by plumbing leaks, the fungus will emerge from the interior of a wooden beam and encase the surface in a white cobweb of mycelium. Once this platform is established, the fungus can use hyphal strands to bridge gaps between beams. The strands can also span concrete or plastics that offer nothing edible to the fungus, allowing the rot to jump from meal to meal, continually nurtured by water piped from the yard by the rhizomorph. This mycological nightmare can keep getting worse. A single, well-established mycelium can also colonize other locations by developing additional indoor rhizomorphs to transfer water, and thereby set up areas of decay all over a house. *Meruliporia* and *Serpula* crawl upward from the basement, issuing between floorboards or cracks in walls, spreading like pancake mix over the home. A homeowner quoted by Kastner said that her infestation looked like Camembert cheese, which is an appropriate description because the Camembert rind is a living mat of hyphae, just like a dry rot colony. In a dry home, every area of decay is dependent on the original umbilical connection to the yard provided by an invading rhizomorph. If this cable can be found and cut, the entire mega-fungus will wither. In a wet home, the dry rot fungus will flourish along with indoor molds, and the original connection to the outdoors may be severed without consequence. In these situations, the fungus can also spread via spores, which brings me to dry rot fruiting bodies.

In a basement suffering from advanced decay, fruiting bodies as big as surfboards burgeon as crusts on the surface of beams and exude an unmistakably fungal smell. These excrescences are yellow to orange-brown in color with a white margin, and form masses of rusty spores (plate 9). Spores are formed in quartets on the prongs of specialized cells called basidia, from which they catapult themselves into the air. In this manner, sizeable fruiting bodies can fill the atmosphere of a rotting basement with a mind-boggling number of spores. A single healthy fruiting body the size of this book could shed billions of spores. In basements where there is little air circulation, the spores collect beneath the crusts, covering furniture, boxes, and sundry clutter with a reddish shroud. Fruiting bodies of *Meruliporia* and *Serpula* can extend all the way along beams, creating colossal platforms for spore production. A Belgian company called Alert-Pest measured a single fruiting body that spread over an area of 6 square meters (or 65 square feet) in a property blanketed with a total of 78 square meters (840 square feet) of spore-producing rot!

As they mature, dry rot fruiting bodies develop ridges, then become pitted with tubes as the ridges extend and fuse with one another, and finally sport a honeycombed or toothed appearance. This process of maturation elaborates an ever greater fertile surface for spore production. Older fruiting bodies can also extend short shelves into the air, like those of bracket fungi on diseased trees and decaying logs. Anatomically and functionally, dry rot fruiting bodies are indistinguishable from edible mushrooms, but nobody would serve them at a dinner party.[8]

Meruliporia caused a great deal of destruction in the southeastern United States in the early part of the twentieth century but has only become a serious problem on the West Coast in the last decade. The reason that dry rot is wrecking homes in Santa Barbara and elsewhere in California may be quite straightforward. To begin transforming productive agricultural land into a sprawling housing development, fragrant orchards of fruit and nut trees are obliterated with the aid of bulldozers. Trees are often snapped aside, rather than pulled out whole, leaving a stump and intact roots in the soil, and the dying tissues are swiftly colonized by fungi. Buried wood will keep the fungi occupied for a while, but once they have rendered all the cellulose into sugar, they begin to search for more food by sending out rhizomorphs. The rhizomorphs extend through the soil, around drainage pipes, and into basements. Sometimes they hit the jackpot: the wood frame of a house. When this happens, mycelia sprout from the ends of the rhizomorphs and penetrate their favorite food source. If the basement is dry as a bone, the fungus imports water as necessary, keeping the wood-chewing mycelia soggy. Imported topsoil and wood mulch brought in for landscaping serve as additional sources of the dry rot fungus. Homeowners can worsen the situation by keeping the rhizomorph soaked with lawn sprinklers, and turning on the air conditioner to cool the fungus to its optimum growth temperature of around 23 degrees Celsius (73 degrees Fahrenheit). It would be difficult to design a more perfect situation for *Meruliporia*.

Changing climatic conditions may aid *Meruliporia* in California and elsewhere, particularly in areas that experience an increase in rainfall, because anything that elevates soil moisture is good for wood-rotting fungi. Dry rot tends to be more common in wet years, because a high water table will maintain the *outdoor* water supply for the *indoor* fungus. But dry rot isn't a new phenomenon. In the Old Testament book of Leviticus, in which Moses is instructed on practical issues ranging from kosher methods of animal slaughter and cooking to the various punishments for having sex with animals or one's mother-in-law, there is a lengthy passage that mycologists have interpreted as an action plan for mold remediation. The Bible identifies the nuisance as a "fretting leprosy of the house,"[9] which betrays itself when the homeowner concludes that "It seemeth to me that there is as it were a plague in the house." Once the home-owner admits his or her fears to a priest, an ordained course of action is initiated. First, the priest must carry out an inspection. If he discovers that "the plague be in the walls of the house with hollow strakes [streaks], greenish or reddish," the home must be closed up for a week.[10] If, after the trial period, the "plague be spread in the walls of the house," the suggested remedy includes scraping the interior of the home and dumping "the dust that they scrape off without the city into an unclean place." (Some mycologists have suggested that Leviticus refers to indoor molds, but the color and behavior of the plague is undeniably more suggestive of dry rot or other wood-decay fungi.) As a last resort, the priest must order the destruction of the house and have the timber and mortar carted off to the (increasingly) unclean place beyond the city walls. This edict would have severely limited mold and dry rot claims by the unin-sured in the ancient world. Contemporary building consultants may see some logic in all this, but I'll never be convinced that killing a bird "in an earthen vessel over running water" (the details of the method—shaking or drowning?—are not specified)[11] is an effective approach to certifying that a home is purified.

Dry rot has ravaged the British more than any other people, or perhaps it would be more accurate to say that the British have been most vocal about its offenses. Remember that the culprit for dry rot in Europe is *Serpula lacrymans*, a fungus with a much longer rap sheet than the Californian menace. *Serpula* has lived with the British for so long, and is so common, that home inspectors expect to find it in anything built more than a couple of decades ago. Like *Meruliporia*, *Serpula* prefers cooler temperatures than many indoor molds, perhaps explaining the prevalence of dry rot, and absence of *Stachybotrys*, in Britain. The outdoor biology of *Serpula* is a mystery, because the fungus has never been found growing on tree stumps anywhere in the British Isles. It does show up, however, in forests in India and in central Europe. Beginning in 1929, an Indian mycologist called Bagchee spent more than 20 years looking for the fungus in the western Himalayas.[12] He found it on tree stumps and fallen logs, and also in buildings. Fruiting bodies developed on the shaded underside of logs in contact with soil, echoing their appearance in basements. The same species was also reported in the 1970s, emerging from stone walls. Might dry rot have moved from India to Britain? In 1992, a multinational expedition relocated *Serpula* in the Himalayas.[13] These investigators suggested that the fungus might have been brought home on the possessions of Victorians returning from India, and then spread to other countries. British imports of Himalayan timber between 1850 and 1920 might have provided an additional shot of dry rot, particularly because the logs were floated down the Ganges to Calcutta and exported without proper drying and treatment. This Indian origin is quite compelling, but it isn't tenable because the British were cursed by dry rot a long time before the raj.

Beyond India, fruiting bodies of *Serpula lacrymans* have been found on the roots of damaged Norway spruce trees in the Czech Republic, suggesting that the fungus sometimes survives as a wound parasite in nature.[14] But many of the sightings of the fungus in Central Europe since the nineteenth century are cases of mistaken identity. The "true" dry rot fungus is easily confused with a related beast called *Serpula himantioides*[15] and is as rare in European forests as it is in India. It is interesting that both *Stachybotrys* and the dry rot fungus torment homeowners but seem to be weak competitors against closely related fungi in the wild (explaining their rarity). Home construction has been highly beneficial to the biological success of these fungi. The mystery of dry rot's immigration to Britain is unlikely to be solved now. Centuries ago, infected timber must have been used to build someone a refuge, and the fungus embraced the opportunity to become a plague upon our houses.

Dry rot of buildings has been tolerated as a fact of life in Britain for centuries and is rarely mentioned by the media. But the fungus was a celebrity in the seventeenth century—like today's toxic black mold—when it ravaged the Royal Navy. As Secretary of Admiralty Affairs in the 1680s, Samuel Pepys surveyed new ships under construction at the Royal Dockyard at Chatham and discovered that "many of them . . . lye in danger of sinking at their very Moorings." The planks were "in many places perished to powder," and "Their Holds not clean'd nor air'd, but . . . suffer'd to heat and moulder, till I have with my own Hands gather'd Toadstools . . . as big as my Fists."[16] *Serpula* wasn't the only fungus attacking the ships at dock. The common woodland species *Polyporus sulphureus*, or sulfur shelf, ate at the oak planks and formed staircases of bright yellow brackets inside the rotting holds. Wood-boring beetles followed the fungi, and their burrows further weakened the timber. (Secondary damage of wood by death watch beetles, woodworm, and other insects is also common in homes.)[17] After assessing "how deeply the Ships were infected with that evil," Pepys concluded that a horrifying proportion of the fleet required complete refitting. This not only led to concern about the cost of the exercise but also about the availability of timber.

Severe timber shortages in Britain began after the Reformation, when the dissolution of the monasteries led to wholesale destruction of their vast wooded estates.[18] Various timber preservation acts were passed, but supplies of serviceable oak trees for shipbuilding continued to diminish. This led to an ever-increasing reliance on timber imported from Eastern Europe, which probably caused the dry rot epidemic. Domestic hardwood tended to be better seasoned—drier and rot resistant—than imported logs, which were often soaked for weeks by floating in rivers, or were rotten on arrival after confinement in the holds of transport ships. The link between the use of unseasoned timber and dry rot was recognized by Pepys,[19] but the need for warships far outstripped the availability of quality lumber. Unseasoned wood was used for repairs to the hull of the *Royal George*, which sank with several hundred crew members at Portsmouth in 1782.[20] A subsequent court martial revealed that the wood had been so rotten that it couldn't hold a nail. In the seventeenth century, warships were expected to last for 25 to 30 years, but dry rot began to severely abbreviate the useful life of a new vessel: By the eighteenth century, the average life of a ship was cut to 12 years, and had dwindled to "no duration" after the Battle of Trafalgar in 1805.[21] In practical terms, ships had became disposable items, which pleased neither crews nor politicians. Nelson's *Victory* and other famous ships from the time of the Napoleonic Wars (1799–1815) were heavily colonized by wood-rotting fungi. The 110-gun battleship *Queen Charlotte* deteriorated so swiftly during construction that the navy was forced to rebuild her in 1810 *before* she could set sail! The financial impact of the damage was alarming, but the government had no choice but to pay.[22] After all, "The royal navy of England hath ever been its greatest defense and ornament; it is its ancient and natural strength; the floating bulwark of the island."[23]

Michael Faraday became involved in the search for a solution to the dry rot problem and delivered a lecture on the subject to the Royal Institution in 1833.[24] He was impressed with a remedy for dry rot patented by Dublin-born inventor John Howard Kyan in 1832.[25] The method became known as wood-kyanizing. Kyan thought that "the evil might be stopped; that the commencement even might be prevented by the application of corrosive sublimate." Previously, corrosive sublimate, or mercuric chloride, had been used to preserve brain tissue and other delicate body parts for scientific study, and also to purge bookworm from library books. Faraday was concerned that the preventive effect against dry rot would be short-lived. Once "the timber of the vessels which were exposed to the bilgewater, and other water, where vessels were not coppered," he wondered, "was the sublimate not likely to be removed, and its effects be destroyed?" The Admiralty ordered a trial in the "fungus pit" at Woolwich. This was "a pit dug in the [naval] yard, and enclosed by wood on all sides, having a double wooden cover; it was damp of itself, and into this were put various kinds of wood, of which they wished to make a trial."[26] Timber was submerged in a tank filled with the corrosive sublimate, dried, transferred to the pit, and left for 5 years. The experiments showed that the treatment was effective against dry rot, and Faraday concluded, "The process here employed completely, and with certainty, prevents the possibility of the destructive effects of the active principle which Nature employs to cause decomposition and decay. . . . Corrosive sublimate neutralizes this primary element of fermentation...rendering [wood] as indestructible as charred timber."

Faraday recognized, however, that Kyan's solution wasn't perfect: "was there not a fatal injury that might arise from the production of a noxious atmosphere?" This had been a concern for users of kyanized library books, too. The jar of mercuric chloride in my laboratory is labeled with a skull-and-crossbones symbol—it damages the nervous system, heart, and other organs, and can be fatal if absorbed through the skin—and a graphic of a dead fish beside a dead tree. Answering his own question, Faraday concluded that once it combined with the treated wood, corrosive sublimate wouldn't pose a serious problem. But his confidence was qualified: "it would be found useful in a far higher degree, in the construction of cottages and outhouses, than palaces; for it is of far more importance to those whose means are small, that they should have that duration given to their timber which would extend the application of their means, and give permanency to their comforts." (To which he should have added, "even if they forget who they are and their hair falls out.") The method wasn't embraced to any great degree by the navy or the public.

A tremendously successful and less noxious preventive for wood decay was introduced in 1838.[27] Creosote is a blend of thousands of compounds derived from coal tar and is effective at stalling the growth of all wood-decay fungi. It seemed like the perfect remedy for galleon-rot, especially when the alternative wood treatments were so obviously incompatible with human health. Some of the constituents of creosote are very nasty chemicals, including trimethylbenzene, naphthalene, and pyrene, but once the wood is impregnated and coated with the stuff, the only danger comes from skin contact. (It is interesting to consider that many of the same cancer-causing chemicals found in creosote are supplied by cigarette smoke.) Besides its immediate application in shipbuilding, creosote proved an impressive defense against rot in fence posts, railroad ties, and telephone poles.[28] Soil and groundwater contamination around treatment plants was problematic, however; after a century and a half of use, the European Union introduced a wholesale ban creosote in 2003. Creosote advocates argue that a ban would lead to a new epidemic of wood decay, and the debate about creosote continues in the United States where the EPA has yet to make a final ruling on the issue.[29] Extermination companies would be happy to see the elimination of creosote, and other businesses also view the potential opportunity for house-destroying fungi as a blessing. Commenting on trends in home construction, an article in *U.S. News and World Report*[30] gave the following plug for the steel industry: "Want a house that's immune to dry rot, moisture, and termites, and resistant to hurricanes and earthquakes? Hire a steel house framer."

Today's steel armadas are resistant to dry rot, but fungi continue to attack wood-hulled pleasure boats. Freeway galleons, otherwise known as RVs, are another target. If a wood-decay fungus can become established in a wet location within the vehicle—flooring beneath a plumbing leak is a likely starting place—it can use its rhizomorphs to colonize the rest of the structure. But stationary homes remain the staple food for dry rot fungi. Insurance companies have used the term "dry rot" in a very general sense to describe the damage caused by any of the wood-decay fungi that can appear in homes, and a blanket exclusion for dry rot is written into most homeowners policies. Nevertheless, some victims of *Meruliporia* have challenged this rule in court by arguing that the damage caused by this fungus represents a new and unforeseen disaster deserving special attention. The California home of Joseph and Jodeanna Glaviano was severely damaged by *Meruliporia*, but the couple were blissfully unaware of this until their hardwood floor collapsed, revealing fungal destruction of the subflooring. Arguing the distinction between dry rot and dry rot caused by *Meruliporia*, attorneys representing the Glavianos were successful in prying some compensation from Allstate Insurance in California.[31] The appeals court ruled that the losses due to "collapse caused by hidden decay" were covered.

The only good thing to say about *Meruliporia* and *Serpula* is that their spores don't carry toxins, so they can't be blamed for lung bleeding or brain damage. This isn't a comforting thing to say to a neighbor confronted with the imminent collapse of their dream home, but can serve as a useful bone to be thrown by a contractor (along with the repair estimate). Dry rot spores are certainly allergenic, and their tremendous numbers in contaminated homes is a concern for anyone exposed to them. Besides asthma, dry rot spores are known to have caused hypersensitive pneumonitis—the immune response involving gamma globulin discussed in chapter 3. A case history published in 1978 described a 32-year-old schoolteacher who developed severe breathing difficulties, joint pains, and began losing weight.[32] His condition improved during a brief spell in hospital, but worsened again when he returned home. Blood tests showed that he had formed high levels of antibodies (IgE and IgG) to *Serpula* spores, which wasn't surprising because his home suffered from extensive dry rot. The relationship between dry rot and allergic disease leaves open the possibility of future personal injury lawsuits as add-ons to claims for property damage.

As a metaphor, dry rot is used frequently as a reflection of human misery, self-imposed or otherwise. In *Great Expectations*, rot served to garnish Miss Havisham's monument to the misfortune of her betrayal at the altar. This queen of self-pity presided over the biodegradation of her uncelebrated wedding banquet, when an afternoon's work with rubber gloves and trash cans would have allowed the woman to reclaim her life. In Poe's *Fall of the House of Usher*, dry rot is promoted to a major character. The narrator of the story describes Usher's mansion in the following manner:

> The discoloration of ages had been great. Minute fungi overspread the whole exterior, hanging in a fine tangled webwork from the eaves . . . there was much that reminded me of the specious totality of old wood-work which has rotted for long years in some neglected vault, with no disturbance of breath of the external air. Beyond this indication of extensive decay, however, the fabric gave little token of instability.[33]

The specious nature of rotten wood is familiar to home inspectors. *Meruliporia* and *Serpula* can turn the interior of a plank into dust before anything more than bubbling of paintwork is visible. Once the cellulose is extracted by the fungus, a screwdriver can be pushed through a 2×4 without any resistance.

Scientific interest centers on the dry rotters, but wet rot fungi that corrode indoor wood that is pre-soaked for the fungus by natural flooding, plumbing disasters, or persistent leaks in the building exterior (or "envelope" in technical parlance) are more common (plate 10). The cellar rot fungus, *Coniophora puteana*, causes a lot of mess in damp basements, and ink cap mushrooms (a species of *Coprinus*) are common on sodden door and window frames. Other fungi are content to blemish the outsides of homes and, more importantly, automobiles. The fact that anyone is concerned about this last category of microorganisms is evidence of a society with little else to worry about. Having pondered this, I have to say that, as the owner of a new car, I have lately become very suspicious of wood mulch. The enemy of a beautifully enameled finish is called *Sphaerobolus (sfare-ob-bol-us) stellatus*, better known as the artillery fungus. This is a coprophilous or dung-loving fungus, adapted for growth on the partially digested cellulose fibers defecated by herbivores or on sodden wood shavings in gardening mulch. Like other kinds of dung fungi, it faces a severe challenge once it has exhausted its food source: How can I escape the dung heap? This has led to the evolution of a series of intriguing mechanical devices that shoot spores onto grass blades surrounding the dung. *Sphaerobolus* uses a miniature trampoline to eject a single 1-mm-diameter black ball (gleba) filled with spores. There's nothing else like this in the living world.

The operation of the fungal trampoline was first described by Pier Antonio Micheli in his *Nova Plantarum Genera*, published in 1729 (plate 11). The trampoline develops as a sphere on the surface of the dung and cracks open to reveal an inner cup that glistens with fluid and holds the gleba. Held under increasing tension, the inner cup finally everts, propelling the gleba up to 5½ meters from the fruiting body—and setting the record for fungal propulsion. An old, slightly flaccid tennis ball serves as a good model: Form a dish by pressing one half into the other, then watch it flip outward to restore the sphere. The rapid motion of the inner cup of the *Sphaerobolus* fruiting body is audible as a popping sound, as is the impact of the gleba on the lid of a culture dish. I find the details of the mechanism fascinating, but rather than sending you to sleep, refer anyone interested in this miraculous organism to a lovely article written by Terence Ingold.[34] To achieve an arching trajectory for its spore mass, *Sphaerobolus* aims for the sun and hits anything in its path, including automobiles. This is a problem, because the glebal mass is very sticky and as it dries becomes glued to the paintwork. Vigorous rubbing will dislodge the black blob, but a circular depression is left in the paint, deep enough that it cannot be buffed out by polishing with a cloth.[35] Automobiles parked next to thousands of the trampolines in a flower bed become conspicuously "dalmatianed," potentially ruining the day for the car owner, the owner's insurance company, mulch supplier, landscaping company, and owner of the parking lot (while brightening the day for the local automobile body shop). One study at Pennsylvania State University estimated that $1 million in claims for automobile damage were filed in a single year in Pennsylvania.[36] Ohio is another hot spot for *Sphaerobolus*, but it grows in most parts of the United States during periods of warm wet weather. House siding is another common target for the fungus, though it's difficult to imagine a home exterior so spotless that it would seem marred by the artillery fungus.

By hitting a car, the fungus is condemned to travel thousands of miles with no prospect of germination.[37] The gleba remains fastened to the vehicle indefinitely, and the spores inside can survive for years with little opportunity to settle on wet mulch or animal dung where they can raise their own families of miniature trampolines. Two centuries ago, the artillery fungus traveled across oceans on the sopping wood of decaying warships. It was figured by the botanical artist James Sowerby during his investigation of the disintegrating *Queen Charlotte* in 1812.[38] Shooting from ship to ship was far more effective than lodging on painted metal, because the fungus had some prospect of continuing its life cycle at sea, and one day of making it back to land when the timber was recycled. This may have enabled *Sphaerobolus* to spread far beyond its prehistoric distribution.

I watched the director's cut of the 1982 movie *Blade Runner* the other evening, and was astonished by the realization that 20 years has passed since I first saw this beautiful film. Milton referred to time as "that subtle thief of youth," and since I'm no longer a youth and yet somehow neglected the slippage of a couple of decades, I humbly concur. In any case, in the movie, the android played by Rutger Hauer remarks upon the wondrous things he has seen, "things you people wouldn't believe"—and realizes that "All these moments will be lost in time, like tears in rain"[39] (followed by "Time to die"). This science fiction classic made me think of molds. (It doesn't take much to do so.) Russian scientists studying samples taken from the *Mir* space station since the 1980s have documented more than 100 different species of fungi growing in the craft.[40] Cosmonauts and astronauts have also noticed that fungal colonies have been growing on the portholes of the International Space Station, prompting extreme efforts to sterilize incoming cargo, and aggressive housekeeping in space. The fungi on the International Space

Station are the same species that grow in homes on Earth, though *Stachybotrys* and *Meruliporia* haven't been found up there yet. After millions of years of unwitnessed toil, the biological careers of fungi happen to have intersected with ours. They followed us from the woods, joined us as cabin mates across oceans and into orbit, and have likely journeyed fantastic distances in unmanned spacecraft. When human history comes to a close, a deluge of spores will help erase the record of our presence on this planet. With a great deal of luck, involving some large and resinous trees, perhaps a mycologist or a mold attorney will be fossilized in amber. Everything else will be lost, like tears in rain.

Notes

CHAPTER 1. *Stachybotrys* versus Superpower

1. Using the same browsers, "fungi" generates close to 1 million hits, and "mold" produces a list of millions of sites (though many are concerned with home decorating, casting, baking, and the town of Mold in Wales).
2. The company was Farmers Insurance, whose experience with mold claims is explored in chapter 6.
3. A. O'Neill, *Los Angeles Times* (April 10, 2002).
4. I take some liberty here in assuming the job titles of Ed's domestics. His Mediterranean-style house (I assume he has others) only has six bedrooms, which offsets my impression that he lives like Marie-Antoinette. While the mold problems were fixed, the McMahons rented another house for $23,000 a month.
5. McMahon's insurers and remediation companies agreed to a $7 million settlement in May 2003.
6. M. D. Clark, *The Cincinnati Enquirer* (June 3, 2002). The title of this article was provocative: "Builder sued after mold evicts woman."
7. P. O'Farrell, *The Cincinnati Enquirer* (June 24, 2001).
8. M. D. Clark, *The Cincinnati Enquirer* (August 27, 2001). The teachers' lawsuit was dismissed later in 2001.

9. W. A. Croft, B. B. Jarvis, and C. S. Yatawara, *Atmospheric Environment* 20, 549–552 (1986).

10. The number of Cleveland infants and the dates of their hospital admission are taken from the review of cases published by Dearborn et al., *Pediatrics* 110, 627–637 (2002).

11. E. Montaña et al., *Pediatrics* 99, 117–124 (1997).

12. Ideally, an inspection expert or firm determines the extent of a mold problem in a building and recommends a course of action (remediation plan). The cleanup is then accomplished by an independent contractor, and finally, the original inspectors (or a third company) return to the site to certify that the problem has been fixed. The plan for removing mold must be supported by a strategy for eliminating the moisture that resulted in mold growth. Otherwise, molds will flourish in the sanitized building as soon as it becomes sufficiently damp.

13. The National Allergy Bureau (NAB) compiles spore and pollen counts on its web site: http://www.aaaai.org/nab/pollen.stm. Volunteers staff 84 counting stations in the United States. Spore counts in excess of 50,000 are ranked in the very high category on the NAB's scale. The highest spore count on record was recorded in 1971 near Cardiff in Wales where 200 million spores per cubic meter were blowing around in the air. That's about three spores per person in the whole of the United Kingdom, which is an exceptionally useless piece of information, like other entries in *The Guinness Book of Records* (New York: Facts on File, 1992).

14. In addition to liberating spores, the possibility that static colonies can release volatile toxins is favored by a few mycologists (and discounted by others).

15. These are the criteria used by a number of industrial hygienists, which serve as a rough guide for building assessment. The square footage refers to the size of individual areas of contamination rather than the sum of numerous small patches of mold growth. A large continuous area suggests significant moisture in this location.

16. Fungi are sensitive to poisoning by metals.

17. See http://www.epa.gov/iaq/pubs/ozonegen.html.

18. Reputable industrial hygienists sometimes use dry ice to freeze mold from wet patches surrounding nails. Freezing the walls and ceiling of a room is a different proposition.

19. The half-life of the isotope cobalt-60 is 5 years, which means that it would take 35 years for the radioactivity of a metallic block of the stuff to decay to less than 1 percent of its starting value.

20. C. H. Tepper, *The Austin Review* (October 31, 2002).

21. *The State of Texas v. Richard L. Steffan et al.*, No. 2002-43363 (Texas

Dist. Ct., 151st Jud. Dist., Harris Cty.); J. A. Zuniga, *The Houston Chronicle* (September 19, 2002).

22. According to David Aurora, in *Mushrooms Demystified: A Comprehensive Guide to the Fleshy Fungi*, 2nd ed. (Berkeley: Ten Speed Press, 1996), the domicile or domestic cup fungus is a nasty thing that appears on plaster, cement, wood, and carpets, and is commonly encountered in shower stalls and damp closets. He also describes its appearance above a bathtub, and in two cars that he has owned. Coincidence seems unlikely here—Aurora must soak the carpeting in his cars on a regular basis.

23. H. J. Chao et al., *Mycopathologia* 154, 93–106 (2002).

CHAPTER 2. Uninvited Guests

1. The manufacturer of a popular wood stain was sued in a class action lawsuit in California for $160 million because the sealant promoted rather than prevented mold growth. Linseed oil is a component of many wood stains. It is thought to stimulate mold growth unless a sufficient quantity of a fungal retardant is also incorporated into the formulation. The story was reported by Barbara Clements, *The News Tribune* (October 31, 2002).

2. "Mould" is the British spelling.

3. House Bill 5040 was introduced by U.S. Representative John Conyers, a Democratic congressman from Detroit. This is discussed in chapter 6.

4. The Romans invented a god of mildew, called Robigus, who was charged with averting crop diseases. Strictly speaking, this deity was supposed to deal with diseases caused by a different group of fungi called rusts, but mycological knowledge was limited in ancient Italy. To rebuff crop damage, worshippers sacrificed a rust-colored dog (and a sheep, for good measure) at the annual festival of Robigalia on April 25. Based on this primeval wisdom, we might try executing someone running a fever to offset global warming.

5. The oomycete water molds constitute a fifth group of microbes studied by mycologists. These employ the same ~~hyphal~~ lifestyle as mushroom-forming fungi but are not at all closely related to them in terms of their ancestry. Their existence is a beautiful example of convergent evolution. These water molds deserve their own book: They caused the potato famine, have ravaged forests with sudden oak death, and one virulent species can dissolve facial tissue.

6. Three common indoor molds—*Aspergillus versicolor*, *Cladosporium cladosporoides*, and *Penicillium melinii*—were studied by R. L. Górny

et al., *Atmospheric Environment* 35, 4853–4862 (2001). The experiments on *Stachybotrys* have not been published. Experiments in which toxic spores are deliberately blown into the air raise significant safety concerns and require containment of the entire experimental apparatus in a safety cabinet.

7. R. A. Haugland, S. J. Vesper, and S. M. Harmon, *Mycologia* 93, 54–65 (2001). *Memnoniella* was outed much earlier, without genetic analyses, by Guy Bisby, who speculated that it might be an unusual form of *Stachybotrys;* see G. R. Bisby, *Transactions of the British Mycological Society* 26, 133–143 (1943). He described *Memnoniella* as "a suspiciously rare fungus," which was one reason why he thought it might be a form of *Stachybotrys*.

8. R. A. Haugland, S. J. Vesper, and L. J. Wymer, *Molecular and Cellular Probes* 13, 329–340 (1999).

9. Although *Stachybotrys* was isolated from root lesions by S. Li et al., *Mycopathologia* 154, 41–49 (2001), this does not prove that the mold is a plant pathogen. It may colonize tissues damaged by other microorganisms.

10. A. D. Phillott et al., *Australian Journal of Zoology* 50, 687–695 (2002). The researchers isolated a species of *Stachybotrys* (not necessarily the same species as the indoor mold), along with other fungi, from the cloaca of female turtles. It is thought that the molds are withdrawn into the body of the reptiles as they retract their lengthy ovipositors from the sand. Experiments on cloacal fungi are motivated by interest in the gynecological health of these endangered animals.

11. D. M. Kuhn and M. A. Ghannoum, *Clinical Microbiology Reviews* 16, 144–172 (2003).

12. B. Andersen, K. F. Nielsen, and B. B. Jarvis, *Mycologia* 94, 392–403 (2002); M. Cruse et al., *Mycologia* 94, 814–822 (2002); and J. Peltola et al., *Canadian Journal of Microbiology* 48, 1017–1029 (2002).

13. D. L. Hawksworth, *Mycological Research* 105, 1422–1432 (2001).

14. The actual number of fungi is a controversial issue. Some scientists have estimated that 2.7 million fungi exist! At the current rate of discovery of 1,000 new fungi per year, it could take 2,000 years to complete the inventory.

15. M. B. Ellis, *Dematiaceous Hyphomycetes* (Wallingford, U.K.: Commonwealth Mycological Institute, 1971); and M. B. Ellis, *More Dematiaceous Hyphomycetes* (Wallingford, U.K.: Commonwealth Mycological Institute, 1976).

16. E. Schuster et al., *Applied Microbiology and Biotechnology* 59, 426–435 (2002).

17. M. R. Tansey and T. D. Brock, *Nature* 242, 202–203 (1973). The fungus is also known as *Dactylaria gallopava*. Besides infecting immunocompromised humans and turkeys, this mold also causes fatal encephalitis in broiler chickens, other birds, and cats.

18. J. U. Ponikau et al., *Mayo Clinic Proceedings* 74, 877–884 (1999).

19. J. U. Ponikau et al., *Journal of Allergy and Clinical Immunology* 110, 862–866 (2002).

20. M. Wainwright, in *Stress Tolerance in Fungi*, edited by D. H. Jennings (New York: Marcel Dekker, 1992), 127–144.

21. Other kinds of fungi have two additional mechanisms for surviving on rocks. They cohabit with photosynthetic algae or bacteria in lichens, or connect with the roots of trees and shrubs growing in the surrounding soil, forming mycorrhizae.

22. K. Sterflinger, *Geomicrobiology Journal* 17, 97–124 (2000).

CHAPTER 3. Carpet Monsters

1. Since 1980, Intal products have been promoted by Fujisawa-Fisons Company, Ltd., based in Osaka (a joint venture between Fisons, U.K. and Fujisawa Pharmaceuticals). Further complicating matters, Fisons no longer exists: it was absorbed by Rhône-Poulenc Rorer in 1995.

2. M. Zureik et al., *British Medical Journal* 325, 411–414 (2002).

3. The list of bright and famous asthmatics includes Antonio Vivaldi, Ludwig van Beethoven, Charles Dickens, Edith Wharton, Dylan Thomas (whose chain-smoking could not have helped his breathing), and Marcel Proust. Some biographers have derided Proust by describing his allergies as psychosomatic, and only recently have scholars analyzed the influence of his disease on his life and work (rather than vice versa). Proust's asthma contributed to his death from pneumonia at age 52.

4. I'm deliberately simplifying the cellular mechanisms in this passage. For example, in addition to their interactions with B cells, T cells also attack invading microbes directly.

5. Cells called basophils and eosinophils also accumulate at the sites of inflammatory reactions and contribute to the resulting tissue damage. Further details are provided in C. A. Janeway et al., *Immunobiology: The Immune System in Health and Disease*, 5th ed. (New York: Garland, 2001). This textbook provides a superb introduction to the science of immunology. Another book I recommend is Lau-

ren Sompayrac, *How the Immune System Works* (Oxford: Blackwell Science, 1999).

6. Mast cells accumulate in the intestine in response to worm infestation. Mutant mice with a mast cell deficit are unable to combat intestinal worms with the vigor shown by a normal mouse. These mutant mice have similar problems dealing with blood-sucking ticks, demonstrating that the immune response is not limited to the gut.

7. Worldwide, intestinal roundworms infect 1.5 billion people, and hookworms parasitize in 1.3 billion people. The filarial worm that causes elephantiasis afflicts 120 million people, and another worm that causes river blindness (by acting as a vector for *Wolbachia* bacteria) infests 30 million in Africa alone. Schistosomes or blood flukes affect 200 million people and cause at least 1 million deaths per year. See D. W. T. Crompton, *Journal of Parasitology* 85, 379–403 (1999).

8. *Vital and Health Statistics* 10 (1995); *Morbidity and Mortality Weekly Report* 47, *Surveillance Summary* 01 (April 24, 1998).

9. E. von Mutius et al., *American Journal of Respiratory and Critical Care Medicine* 149, 358–364 (1994).

10. S. T. Weiss, *New England Journal of Medicine* 347, 930–931 (2002).

11. D. R. Ownby, C. C. Johnson, and E. L. Peterson, *Journal of the American Medical Association* 288, 963–972 (2002). A related study found that cat lovers with black felines were more likely to suffer from nasal allergy than those with light-colored ones. This could be explained if black cats produced higher concentrations of a potent cat allergen called fel d 1, but this has not been tested.

12. Some allergists would add a pinch of respiratory syncytial virus to the recipe.

13. P. Van Eerdewegh et al., *Nature* 418, 426–430 (2002), and associated commentary by J. M. Drazen and S. T. Weiss, *Nature* 418, 383–384 (2002).

14. C. H. Blackley, *Experimental Researches on the Causes and Nature of Catarhus aestivus (Hay-fever or Hay-asthma)* (London: London, Baillière, Tindall & Cox, 1873).

15. An excellent description of early studies on the allergenic effects of fungal spores is given in G. C. Ainsworth, *Introduction to the History of Mycology* (Cambridge: Cambridge University Press, 1976).

16. Here's the calculation. I based my estimate of 10 spores per minute on figures in *Mr. Bloomfield's Orchard* (chapter 8, note 16): 10 spores per minute × 60 minutes × 24 hours × 365 days × 80 years = 420,480,000 spores per lifetime. I modeled spore mass on a

sphere with radius of 3 millionths of 1 meter (3 micrometers). This cell has a volume of 10^{-16} cubic meters, or 10^{-10} milliliters; because 1 milliliter of water weighs 1 gram, spore weight will equal 10^{-10} grams. Multiplying spore weight by number of spores provides an estimated total weight of 0.04 grams.

17. The few hundredths of a gram of mold spores inhaled in a lifetime make an interesting counterpoint to the amount of tar inhaled by a smoker. A single low-tar cigarette releases about 10 milligrams of tar. The lungs of an individual who smokes one pack a day for 40 years will process a total of 2.9 kilograms, or 6.4 pounds of the carcinogenic treacle. That's enough to fill a 3-quart jug.

18. The method for measuring the amount of beta-glucan is interesting. It relies on the sensitivity to these molecules by cells called amoebocytes in the horseshoe crab, *Limulus polyphemus*. Amoebocytes are the only type of cell in the bloodstream of the animal and cause clot formation when they encounter bacterial toxin or the fungal glucan (evidence of infection). These cells perform the same trick in a test tube, which has created a market for the blue blood of the crab. Fortunately, blood can be collected using a syringe without permanent injury to the animal.

19. G. D. Brown and S. Gordon, *Nature* 413, 36–37 (2001).

20. The following book is a superb resource for information on fungal allergens: M. Breitenbach, R. Crameri, and S. B. Lehrer, eds., *Fungal Allergy and Pathogenicity* (Basel: Karger, 2002).

21. C. Barnes et al., *Annals of Allergy, Asthma, and Immunology* 89, 29–33 (2002). Antibodies that bind to a *Stachybotrys* protein may also recognize an allergen produced by another mold, raising the possibility that the antibodies identified in this study may have been generated in response to exposure to other mold species. This point was made by Jon Musmand (a physician from Maine) in a letter published in *Annals of Allergy, Asthma, and Immunology* 90, 274–275 (2003) and conceded by Charles Barnes (his response appeared in the same issue of the journal on pages 275–276). In his letter, Musmand cites studies by other investigators who have failed to detect antibodies (IgE) against *Stachybotrys*, even among people exposed to the fungus.

22. This unpublished research on *Stachybotrys* proteases was performed by Iwona Yike and Dorr Dearborn in Cleveland, and their colleague Thomas Rand at St. Mary's University in Nova Scotia.

23. In addition to allergic responses to large quantities of spores and other allergenic materials inhaled from plant materials, a distinct influenza-like illness may be caused by toxins associated with the

dust (e.g., mycotoxins on the spores). This illness has been referred to as pulmonary mycotoxicosis, grain fever, silo unloader's syndrome, and organic dust toxic syndrome. A case involving employees at a golf course who had unloaded moldy wood chips from a trailer was reported by the CDC: *Mortality and Morbidity Weekly Report* 35, 483–484, 489–490 (1986).

24. M. Jaakola et al., *Clinical and Experimental Immunology* 129, 107–112 (2002). Maritta Jaakkola and colleagues in Finland have found *Stachybotrys*-specific IgG among adults but conclude that this signature of mold exposure does not correspond to an increased incidence of asthma. (Adult-onset asthma was correlated, however, with the formation of antibodies specific for another indoor mold called *Trichoderma citrinoviride*.) I don't think anyone would disagree with the following statement by Charles Barnes: "Whether or not *Stachybotrys* allergy will become a proven cause of allergic disease must await further experimentation."

25. I. Yike et al., *Mycopathologia* 154, 139–152 (2002); I. Yike et al., *Mycopathologia* 156, 67–75 (2002).

26. Expulsion of as much blood as possible allows the investigator to obtain clear sections of tissues for histological analysis.

27. We must be cautious in extrapolating the development of *Stachybotrys* inside the lungs of rat pups to human infants or adults. The method of exposure was anything but natural in the animal experiments, and rat lungs are tiny things that do not offer a perfect model of the human version. Rat pups that were allowed to recover were able to clear germinated spores from their lungs in a few days. Experiments on adult animals demonstrate that the spores of other molds can be expelled from the lung within minutes, so it's possible that the slow clearance of *Stachybotrys* spores was due to the youth of the animals.

28. The issue of spore size and inhalation is discussed by R. V. Miller, C. Martinez-Miller, and V. Bolin, in *Mycotoxins and Phycotoxins in Perspective at the Turn of the Millennium*, edited by W. J. de Koe et al. (Wageningen, The Netherlands: W. J. de Koe, 2001), 123–131; Yike et al., *Mycopathologia* 154, 139–152 (2002); and D. M. Kuhn and M. A. Ghannoum, *Clinical Microbiology Reviews* 16, 144–172 (2003). M. Geiser et al., *Journal of Allergy and Clinical Immunology* 106, 92–100 (2000), provide clear evidence for alveolar deposition of puffball spores with a diameter of 3.1 micrometers.

29. The suggestion that *Stachybotrys* is a newcomer in American homes is discussed in chapter 7.

30. M. Seuri et al., *Indoor Air* 10, 138–145 (2000). *Sporobolomyces* is a

basidiomycete fungus. The levels of the spores formed by mushrooms and other basidiomycete fungi in outdoor air have a strong effect on the severity of asthma symptoms in children. This is discussed by E. Levetin and W. E. Horner, in *Fungal Allergy and Pathogenicity*, edited by M. Breitenbach, R. Crameri, and S. B. Lehrer (Basel: Karger, 2002), 10–27.

CHAPTER 4. Mycological Warfare

1. N. P. Money, *Mr. Bloomfield's Orchard: The Mysterious World of Mushrooms, Molds, and Mycologists* (New York: Oxford University Press, 2002).
2. J. W. Bennett and M. Klich, *Clinical Microbiology Reviews* 16, 497–516 (2003).
3. K. F. Nielsen, *Fungal Genetics and Biology* 39, 103–117 (2003).
4. See chapter 2, note 12.
5. The abbreviation for "variety" is "var." In this case, the scientists designated var. *jateli* as a particularly toxic strain of the mold they called *Stachybotrys alternans*. In the absence of the fungus cultured in the 1940s, the name *Stachybotrys alternans* var. *jateli* is considered equivalent to *Stachybotrys chartarum*.
6. J. Forgacs, in *Microbial Toxins. Volume VIII: Fungal Toxins*, edited by S. Kadis, A. Ciegler, and S. J. Ajl (New York: Academic Press, 1972), 95–128. I drew on this and other review articles by Forgacs for much of the information on stachybotryotoxicosis in the former Soviet Union and Eastern Europe.
7. A. Kh. Sarkisov and V. N. Orshanskayia, *Veterinariya* 21, 38–40 (1944).
8. Hyphal growth within the lungs of rat pups was discussed in chapter 3 (see chapter 3, note 25).
9. B. Freedman, *McIlvania* 14, 83–87 (1999). Stachybotryotoxicosis was one of many disasters in Ukraine in the 1930s. Millions of Ukrainians died during the famine resulting from Stalin's forced collectivization of farms. Khrushchev describes the way that he handled the outbreak in his autobiography, *Khrushchev Remembers* (Boston: Little, Brown, 1970), 110–114.
10. The study from Stanford was J. R. Bamburg and F. M. Strong, in *Microbial Toxins. Volume VII: Algal and Fungal Toxins*, edited by S. Kadis, A. Ciegler, and S. J. Ajl (New York: Academic Press, 1971), 207–292. Trichothecene synthesis by *Stachybotrys* was reported by R. M. Eppley and W. J. Bailey, *Science* 181, 758–760 (1977).

11. If you do not recognize this animal, go to http://www.sea-monkey
    .com. Animated by adding water to their eggs, brine shrimp are
    "The World's Only Instant Pet." They are very interesting things to
    watch, but I'm sure that I was not the only child to be disap-
    pointed to learn that sea monkeys are just a few millimeters long. I
    rank this experience with the tragic realization that X-ray specs do
    not allow one to see through clothing.
12. I've moved on from Michele Pfeiffer (see *Mr. Bloomfield's Orchard*,
    page 22).
13. I. Yike et al., *Applied and Environmental Microbiology* 65, 88–94
    (1999).
14. G.-H. Yang et al., *Toxicology and Applied Pharmacology* 164, 149–160
    (2000).
15. A recent study showed that mushrooms engage in apoptosis to
    avoid releasing faulty spores: B. C. Lu, N. Gallo, and U. Kües, *Fun-
    gal Genetics and Biology* 39, 82–93 (2003); N. P. Money, *Nature* 423,
    26 (2003).
16. *Stachybotrys* and other black molds do not produce vomitoxin.
    This trichothecene is characteristic of species of the fungus *Fusar-
    ium*, and vomiting is a symptom of poisoning following the con-
    sumption of *Fusarium*-contaminated corn. The bioassay for vomi-
    toxin is remarkable: Day-old ducklings are injected with the
    compounds and watched until they begin to vomit at 2- to 3-
    minute intervals.
17. S. J. Vesper et al., *Applied and Environmental Microbiology* 65, 3175–
    3181 (1999); citations of later studies are given by S. J. Vesper and
    M. J. Vesper, *Infection and Immunity* 70, 2065–2069 (2002).
18. L. Gregory et al., *Mycopathologia* 156, 109–117 (2002).
19. J. M. Van Emon et al., *Journal of Occupational and Environmental
    Medicine* 45, 582–591 (2003). The researchers report a serum stachy-
    lysin concentration of 0.4 micrograms per milliliter in blood pooled
    from five occupants of buildings contaminated with *Stachybotrys*.
    This estimate is difficult to square with the amount of stachylysin
    likely to be carried by spores (and assumes highly efficient absorp-
    tion of the toxin by lung capillaries). A single spore of *Stachybotrys*
    weighs roughly 10^{-10} grams (0.1 nanograms; see chapter 3, note 16).
    Even if the stachylysin content of each spore equaled 10 percent of
    its weight (0.01 nanograms), the absorption of all of the stachylysin
    carried by 100 million spores would be required to produce the es-
    timated serum concentration of 0.4 micrograms per milliliter (based
    on an adult serum volume of 2,500 milliliters). The investigators
    discussed the possibility of long-term accumulation of the toxin,
    but this level of spore inhalation is impossible for anyone. Never-

theless, their identification of stachylysin in human serum is a very significant finding.

20. H. A. Burge, in *Indoor Air and Human Health*, 2nd ed., edited by R. B. Gammage and B. A. Berven (Boca Raton, Fla.: CRC Press, 1996), 171–178.

21. R. V. Miller, C. Martinez-Miller, and V. Bolin, in *Mycotoxins and Phycotoxins in Perspective at the Turn of the Millennium*, edited by W. J. de Koe et al. (Wageningen, The Netherlands: W. J. de Koe, 2001), 123–131.

22. R. L. Górny et al., *Applied and Environmental Microbiology* 68, 3522–3531 (2002).

23. G. S. Sidhu, *European Journal of Plant Pathology* 108, 705–711 (2002).

24. Rand reported a reduction in collagen distribution in the lungs of juvenile mice instilled with *Stachybotrys* spores: T. G. Rand et al., *Mycopathologia* 156, 119–131 (2003). A similar result was also suggested by the pattern of lung damage seen in rat pup experiments discussed in the previous chapter (chapter 3, note 22).

25. T. Kordula et al., *Infection and Immunity* 70, 419–421 (2002). In this study, stachyrase A was purified from hyphae growing in a nutrient broth, which means that we cannot be sure that this enzyme is carried by spores.

26. See Money, *Mr. Bloomfield's Orchard* (2002).

27. Nielsen (2003).

28. R. W. Wannemacher and S. L. Wiener, in *USAMRICD Textbook of Military Medicine. Medical Aspects of Chemical and Biological Warfare*, edited by F. R. Sidell, E. T. Takafuji, and D. R. Franz (Washington, DC: Office of the Surgeon General at TMM Publications, 1997), 655–676. This is a superbly written literature review, a true diamond in the rough of government publications.

29. Bamburg and Strong (1971).

30. German toxicologist Manfred Gareis has discussed this unpublished case history at meetings.

31. See Wannemacher and Wiener, *USAMRICD* (1997).

32. T. D. Seeley et al., *Scientific American* 253, 128–137 (1985). A good read for a rainy afternoon.

33. Wannemacher and Wiener, *USAMRICD* (1997).

34. Two strains of *Stachybotrys chartarum* obtained from bees are maintained in the Microfungus Collection and Herbarium at the University of Alberta.

35. Forgacs (1972). In addition to the cases of human stachybotyotoxicosis in the former Soviet Union, exposure to animal feed contaminated with *Stachybotrys* has sickened people in other European countries including Croatia, Bosnia and Herzegovina, Yugoslavia,

and Hungary. Hungarians exposed to contaminated straw suffered nose bleeding. See Andrássy et al., *Mykosen* 23, 130–133 (1980); Y. Ueno, editor, *Trichothecenes: Chemical, Biological, and Toxicological Aspects* (New York: Elsevier, 1983).

36. A. Z. Joffe, in *Microbial Toxins. Volume VII: Algal and Fungal Toxins*, edited by S. Kadis, A. Ciegler, and S. J. Ajl (New York: Academic Press, 1971), 139–189.

37. The policy information was obtained from the American Type Culture Collection, Manassas, Virginia (http://www.atcc.org). Different rules apply to pathogenic fungi, and a number of countries are barred from ordering cultures of these organisms. A handful of dangerous fungal species require containment and handling in the laboratory according to regulations for Biosafety Level 3—the same as anthrax, HIV, and rabies. The customer accepts responsibility for safety once the cultures are received. *Coccidioides immitis* is one of these fungi, which causes a potentially fatal disease called valley fever in the southwestern United States and Mexico. *Histoplasma capsulatum*, the cause of Ohio Valley disease, is another.

38. H. R. Burmeister, *Applied Microbiology* 21, 739–742 (1971). The study concerned T-2 production by *Fusarium tricinctum*, but similar methods would certainly yield different kinds of trichothecenes from other molds. Note to arms inspectors: Rumpled photocopies of this paper in a lab are a good indication that somebody has been misbehaving.

39. Experiments show that mice are far more sensitive to trichothecene poisoning when the toxins are inhaled rather than injected. D. A. Creasia et al., *Fundamental and Applied Toxicology* 8, 230–235 (1987).

40. Wannemacher and Wiener, *USAMRICD* (1997).

41. R. A. Zilinskas, *JAMA* 278, 418–242 (1997). Zilinskas participated in the UNSCOM investigation of Iraq's weapons of mass destruction in the 1990s.

42. According to the German news magazine *Der Spiegel* (January 30, 1989, 16–18), the purchase of trichothecenes from Sigma was brokered by Josef Kuhn, head of a German export firm and suspected Iraqi agent.

43. Jonathan Tucker is the author of *Scourge: The Once and Future Threat of Smallpox* (Boston: Atlantic Monthly Press, 2001).

CHAPTER 5. Cleveland Revisited

1. Pulmonary hemosiderosis can be caused by cow milk allergy (sometimes known as Heiner's syndrome); another cause is Goodpasture's

syndrome, a disease accompanied by kidney damage that affects young adult males.

2. The American Academy of Pediatrics began recommending back sleeping in 1992. Since then, the frequency of prone sleeping has been slashed, and the number of SIDS cases has decreased by 40 percent. See *Pediatrics* 105, 650–656 (2000).

3. Sure it's a stereotype, but go look at a symphony audience for proof that business is thriving for American furriers and plastic surgeons.

4. D. Davis and E. Marchak, *The Plain Dealer* (March 12, 2000).

5. Many articles in the scientific literature refer to 9 or 10 cases of idiopathic pulmonary hemosiderosis (IPH) treated at Rainbow in 1993 and 1994. These numbers reflect the earliest reports of the IPH cluster that appeared after Dearborn enlisted the help of the CDC (the CDC became involved after Rainbow had admitted its eighth case). The numbers quoted in this chapter are taken from a review of cases published by Dearborn and colleagues in *Pediatrics* 110, 627–637 (2002). Eight infants were admitted between January 1993 and November 1994, and another three cases were treated in December 1994. An additional patient described in the article was the twin sister of a boy treated for IPH; though she was not admitted to Rainbow for lung hemorrhaging, subsequent evaluation suggested that she had also suffered from lung bleeding. Some inconsistencies remain, however, between the timeline established in the *Pediatrics* paper and other sources. Robyn Meredith reporting for the *New York Times* (January 24, 1997) described four cases of IPH admitted to Rainbow during two days in November 1994, but the review in *Pediatrics* details the treatment of a single baby for the whole month (a boy who was readmitted three weeks after his first episode of lung bleeding). The *New York Times* article also reported a 1,000-fold increase in the number of IPH cases in Cleveland; although this is true, it is important to recognize that the estimated increase was based on a highly unusual but tiny cluster of initial cases. Dearborn is quoted by the *New York Times* as saying, "Six came in with nosebleeds. Three of them were dead within a week." The *Pediatrics* review refers to three deaths among the 21 babies admitted to Rainbow preceding publication of the *New York Times* article (two more deaths occurred between 1997 and 1999), consistent with a more modest 14 percent mortality rate than the 50 percent figure implied by the quote. A compassionate reader might say that the journalist was guilty of overexuberance in representing the magnitude of the outbreak.

6. E. Montaña et al., *Pediatrics* 99, 117–124 (1997).

7. D. M. O. Becroft and B. K. Lockett, *Pediatrics* 101, 953–954 (1998); Dearborn and colleagues reply to the letter on pages 954–955 of the journal. An analysis of infant deaths in Denver, Colorado, suggested that iron-laden macrophages accumulate in cases of homicidal asphyxia involving repeated smothering of an infant, but not after a single instance of lethal asphyxiation (e.g., by drowning): D. A. Schluckebeir et al., *The American Journal of Forensic Medicine and Pathology* 23, 360–363 (2002).

8. The quote is taken from J. B. Beckwith, *Annals of the New York Academy of Sciences* 533, 37–47 (1988).

9. In addition to Dearborn and Etzel, the initial investigations involved Terry Allan and Timothy Horgen from the Cuyahoga County Board of Health, and Eduardo Montaña from the CDC.

10. The method of science doesn't really allow us to prove anything, rather it simply supports or refutes specific ideas (sometimes codified as formal hypotheses). Frank Harold defined science, with his characteristic crispness, as "The endeavor to understand the universe on the basis of reason, observation, and experiment" (with emphasis placed on *endeavor*). But having said this, scientists are certain enough that our planet revolves around the sun, rather than vice versa, and that humans are a species of African ape. Likewise, evolution by natural selection is as close to a fact as the suggestion that apples fall from trees rather than flying skyward. If someone were to tease out a sliver of evidence suggesting that the assortment of organisms on the planet has not changed over the last billion years, I'll be very surprised (and enroll in a sprint course on Christian fundamentalism to determine how best to avoid cleansing fires, insertion of hot pokers, etc.).

11. C. D. Cassimos, C. Chryssanthopoulos, and C. Panagiotidou, *Journal of Pediatrics* 102, 698–702 (1983).

12. E. Montaña et al. (1997).

13. *Mortality and Morbidity Weekly Report* 46, 33–35 (1997). This report made the first mention of *Stachybotrys* in the Cleveland cases.

14. W. A. Croft, B. B. Jarvis, and C. S. Yatawara, *Atmospheric Environment* 20, 549–552 (1986). This study is also discussed in chapter 1.

15. Michael Sangiacomo reported in *The Plain Dealer* (September 14, 1995) that health officials were "99 percent certain" that *Stachybotrys* caused the lung bleeding cases in Cleveland. Katherine Rizzo wrote an article titled "Baby-Killing Fungus" for the Associated Press wire (July 31, 1997), and *CNN* reported in 1997 that all of the infants in the initial outbreak had died. Anchors on prime-time television news broadcasts showed no restraint, employing their familiar scare-

mongering tactics to hook viewers. In their interviews with journalists, Dorr Dearborn and other investigators revealed their convictions about *Stachybotrys* but didn't say that the case was proven. Dearborn was quoted as saying that "scientists are '90 percent' certain that the fungus is responsible, but . . . that finding the final 10 percent proof would be hard" (M. Sangiacomo, *The Plain Dealer*, February 14, 1996).

16. R. A. Etzel et al., *Archives of Pediatrics and Adolescent Medicine* 152, 757–762 (1998). This paper reported 20 million colony-forming units (CFU or viable spores) per gram of dust sample taken from case homes, versus 7,000 CFU per gram in control homes.

17. In 1999, the insurance industry spent $86 million in lobbying expenses, $5 million less than the top "investment" made by the pharmaceutical industry (source: http://www.opensecrets.org).

18. *Mortality and Morbidity Weekly Report* 49, 180–184 (2000).

19. Ruth Etzel and six coauthors involved in the continuing research in Cleveland refute the criticisms leveled by the CDC panel on this web site: http://gcrc.meds.cwru.edu/stachy/InvestTeamResponse.html.

20. R. A. Etzel et al. (1998).

21. J. C. May, *My House is Killing Me! The Home Guide for Families with Allergies and Asthma* (Baltimore: The John Hopkins University Press, 2001).

22. R. A. Etzel et al. (1998).

23. The quote comes from Harbison's presentation at the Harris-Martin's Mold Litigation Conference held in New Orleans in June 2003.

24. B. B. Jarvis et al., *Applied and Environmental Microbiology* 64, 3620–3625 (1998).

25. S. Vesper et al., *Journal of Urban Health* 77, 68–85 (2000).

26. O. Elidemir et al., *Pediatrics* 102, 964–966 (1999). There is a second instance of *Stachybotrys* isolation from lungs, but this was from a 65-year-old man in Canada.

27. The potential role of mycelial fragments in delivering mycotoxins has been discussed a number of times in the book, but it may be worth underscoring the hypothesis. Because mycotoxins are carried in the outer shell of the spores, any fragments of cell-wall material derived from mold colonies probably carry the same toxins as intact spores. Indeed, the fungus might be dead long before the toxin-bearing fragments are dispersed into the air by disturbance to the colonies. Fragments will not germinate on culture plates, so they could contaminate a room that appears to be free from mold when attempts are made to culture fungi from air samples. The small size

of the fragments relative to whole spores, and their irregular shapes, would also hide these particles from investigators trained to count spores on microscope slides.

28. M. Nikulin et al., *Fundamental and Applied Toxicology* 35, 182–188 (1997).

29. I. Yike et al., *Mycopathologia* 154, 139–152 (2002). An earlier study on rats also found lung damage at high spore dosages: C. Y. Rao, D. B. Joseph and H. A. Burge, *Applied and Environmental Microbiology* 66, 2817–2821 (2000).

30. Despite the criticisms of the original work by the CDC, Dorr Dearborn received funding from the National Institutes of Health to pursue research on the animal model for IPH. Ruth Etzel is not directly involved in this research effort. She was reassigned after the CDC's ruling and later began working for the Southcentral Foundation in Anchorage, a nonprofit corporation that provides healthcare and education services to the Alaskan Native community.

31. *Mortality and Morbidity Weekly Report* 44, 67 & 73–74 (1995).

32. S. M. Flappan et al., *Environmental Health Perspectives* 107, 927–930 (1999).

33. I. Dill, C. Trautmann and R. Szewzyk, *Mycoses* 40, 110–114 (1997).

34. W. G. Sorenson, *Environmental Health Perspectives* 107, 469–472 (1999).

35. Croft, Jarvis, and Yatawara (1986).

36. W. A. Croft et al., *Journal of Environmental Biology*, 23, 301–320 (2002). This paper refers to the idea that, under certain environmental conditions, volatile mycotoxins could be released into the air. Croft works for the Environmental Diagnostic Group, a company in Madison, Wisconsin, and is not affiliated with an academic institution.

37. See http://www.cdc.gov/nceh/airpollution/mold/moldfacts.html.

38. See chapter 4.

39. I. Yike et al., *Mycopathologia* 156, 67–75 (2002).

40. C. Barnes et al., *Annals of Allergy, Asthma, and Immunology* 89, 29–33 (2002). This study is discussed in chapter 3.

CHAPTER 6. Your Verdict, Please

1. *Ballard v. Fire Insurance Exchange, A Member of the Farmers Insurance Group*, No. 99-05252 (Texas Dist. Ct. 345th Jud. Dist., Travis City).

2. Ronald's medical condition was referred to as "toxic encephalopathy," which did nothing to clarify the nature of the illness because

this is a general term that describes any disease of the brain resulting from poisoning.

3. I quote from comments made by Melinda Ballard at HarrisMartin's Mold Litigation Conference held in San Antonio in October 2002.

4. A. Hamilton, *Time* (July 2, 2001).

5. L. Belkin, *The New York Times Magazine* (August 12, 2001).

6. C. Grisales, *The Austin American-Statesman* (October 20, 2002).

7. Losses in the billions of dollars have also been quoted by insurance companies in other states. Some analysts contend, however, that much of these losses were due to the poor performance of the industry's investments in the stock market. By omitting details, investment tribulations since 1999 could be concealed under the blanket of spores because the litigation happened to occur at the same time.

8. Adding to its woes, Farmers Insurance was sued for $150 million in 2002 by state regulators in Texas who claimed that the company was using deceptive trade practices and overcharging customers. In November 2002, the company agreed to pay a $100 million settlement and reversed its decision to withdraw from the residential insurance market.

9. Grisales (October 20, 2002).

10. *Allison v. Fire Insurance Exchange, A Member of the Farmers Insurance Group*, No. 03-01-00717-CV (Texas Ct. App. 3rd Dist., Austin).

11. The power of personality tests to predict vocation is highly questionable, as a high proportion of academics seem better suited to work in a carnival rather than a seat of higher learning.

12. F. X. Finnigan, *HarrisMartin Columns—Mold* (August 2002).

13. There are certainly examples of buildings that become moldy after changes in ventilation, but the cause and effect relationship is usually overstated. For example, Finnish researchers described the management of a mold problem in school buildings that were heavily contaminated after ventilation holes in wood flooring were plugged to save energy costs: U. Haverinen et al., *Environmental Health Perspectives* 107, 509–514 (1999). More importantly, the school buildings also had numerous construction defects, had been built in a very wet location, and had suffered major water damage after a fire.

14. K. McAllister, *Journal News* [Hamilton, Ohio] (February 22, 2003). Animal Planet and the Discovery Channel featured Hunter during training at the Florida Canine Academy. Mold dogs trained by Bill Whitstine are now working in every major city in the United States.

15. The quote comes from a report by Carmel Cafiero broadcast on *7 News*, WSVN-TV in Miami–Ft. Lauderdale, Florida (January 31, 2003).

16. By "techniques" I mean areas of inquiry that demand expertise

with a certain suite of methods. The trend away from mycologists, phycologists, nematologists, and mammalogists is clear from a perusal of job advertisements for biologists. Statistical analysis is unnecessary here; I am confident that few, if any, biologists will disagree with this assessment of the sociological change. John Tyler Bonner discussed the impact of these changes at Princeton University in his autobiography *Lives of a Biologist: Adventures in a Century of Extraordinary Science* (Cambridge, Mass.: Harvard University Press, 2002). The technological sophistication of the methods used by a particular researcher is, of course, no indicator of market interest. My mentor Frank Harold once joked that, because most researchers had retooled their labs to run gels, my insistence on learning things about fungi by poking their cells with glass pipets (which is exceedingly difficult) was no more relevant than the skills of a wheelwright after the introduction of Ford's Model T. In my defense, I note that Frank still spoke of his "radiogram" by the time most Americans owned a CD player.

17. *Daubert v. Merrell Dow Pharmaceuticals* (92–102), 509 U.S. 579 (1993). I quote from the Supreme Court's syllabus (or synopsis). In his opinion on *Daubert*, Supreme Court Chief Justice Rehnquist wrote that the justices were forced to "deal with definitions of scientific knowledge, scientific method, scientific validity, and peer review—in short, matters far afield from the expertise of the judges. . . . I defer to no one in my confidence in federal judges; but I am at a loss to know what is meant when it is said that the scientific status of a theory depends upon its 'falsifiability,' and I suspect that some of them will be, too."

18. A. W. Campbell, *HarrisMartin Columns—Mold* (November 2002). This is not a peer-reviewed article, but if you refer to my discussion of the paper by Croft in chapter 5 (note 36), you'll see that anecdotal studies on mold and health also appear in low-status peer-reviewed journals. A study isn't necessarily authenticated by the process of peer review.

19. W. A. Gordon, in *Bioaerosols, Fungi, and Mycotoxins: Health Effects, Assessment, Prevention, and Control*, edited by E. Johanning (Albany, New York: Eastern New York Occupational and Environmental Health Center, 1999), 94–98.

20. *Linda Maxwell and Rheannon Maxwell, a minor, by her parent Linda Maxwell, v. Pleasonton Unified School District*, No. V-015520 (Calif. Super. Ct., Alameda County).

21. Commenting about endangered species and the obstacle they represent to developers, Miller said, "The endangered are endangered for a reason. They are too dumb to live." If this is true, the omission

of his name from the Federal List of Endangered and Threatened Wildlife is unconscionable. The reason for Miller sharing his analysis of the kangaroo rat at a conference on mold damage was unclear.

22. C. Hutchinson and H. R. Powell, *A New Plague—Mold Litigation: How Junk Science and Hysteria Built an Industry* (Washington D.C.: U.S. Chamber Institute for Legal Reform, 2003), and B. D. Hardin et al., *A Scientific View of the Health Effects of Mold* (Washington D.C.: U.S. Chamber Institute for Legal Reform, 2003).

23. G. Sealey, accessed at http://www.abcnews.com (May 13, 2002).

CHAPTER 7. Everlasting Strife

1. This was the first thing that came to my mind, but a sulfur spring or hydrothermal vent would substitute for this troubling image.

2. A prevalent view of biological diversity places all organisms into one of six kingdoms: eubacteria, archaebacteria, fungi, plants, animals, and protists. Other authorities prefer a larger number of kingdoms, adding, for example, the kingdom stramenopila, which includes oomycete water molds (also referred to as oomycete fungi), brown algae, and diatoms.

3. J. Rikkinen et al., *Mycological Research* 107, 251–256 (2003).

4. There is a vast literature on fungal melanins; the most up-to-date review article on the subject at the time of writing was published by J. D. Nosanchuk and A. Casadevall, *Cellular Microbiology* 5, 203–223 (2003).

5. M. L. Butler et al., *Mycologia* 93, 1–8 (2001).

6. As reported on http://www.news.bbc.co.uk (June 22, 2001).

7. This suggests a good title for a research project: "Antifungal and antibacterial effects of *Stachybotrys chartarum*." It is good because nobody has studied this properly, fairly simple experiments are guaranteed to yield interesting data, and the findings are likely to be of practical significance.

8. Evidence for a succession of molds is reported by E. Pieckova and Z. Jesenska, *Annals of Agricultural and Environmental Medicine* 6, 1–11 (1999). Repeated cycles of appearance-dominance-disappearance of one fungus after another is a well-known phenomenon. Disappearance is due to the dispersal of spores. The succession of fungi on animal dung is the best understood example of this ecological process.

9. I. K. Ross, *Biology of the Fungi: Their Development, Regulation, and Associations* (New York: McGraw Hill, 1979), page 389.

10. I don't expect most of my readers to be interested in this, but I insert this note for my current employers: the taxpayers of Ohio. My working hours are normally spent trying to reformat the hard-drives of Midwestern students, but I qualified for a research leave in 2002 and so hoofed off in search of black mold. I'm back in my shed now, eking out writing time whenever I can between my class-room rants on evolution. It was minus 10 Fahrenheit the other morning, cold enough to coagulate an atheist. This one's feet went numb while it recounted its visit to balmy Santa Barbara.

11. P. J. Davis, *California Research Bureau Note* 8, No. 1 (2002).

12. T. M. Husman, *Environmental Health Perspectives* 107 (Supplement 3), 515–517 (1999).

13. B. Freedman, *McIlvania* 14, 83–87 (1999).

14. In addition to the strains obtained from homes, *Stachybotrys chartarum* has also been isolated from plant pots, moldy sacking, cardboard, sand, feathers, and cat hair. Isolates from the guts of bees were mentioned in chapter 4.

15. In their masterful review article in *Clinical Microbiology Reviews* 16, 144–172, (2003), D. M. Kuhn and M. A. Ghannoum cite studies from the 1980s which showed that 12 percent of buildings in the United Kingdom suffered from serious dampness.

16. Davis (2002).

17. Kuhn and Ghannoum (2003).

18. Climatic conditions have a profound influence on fungal growth and mycotoxin production. Unusual weather is thought to have played a role in the outbreak of alimentary toxic aleukia (ATA) in the 1940s in Russia. (ATA is discussed in chapter 4.) The ATA epidemic was caused by the ingestion of moldy grains that had over-wintered in the fields. Laboratory experiments showed that cultures of the *Fusarium* species responsible for the illness were particularly toxic if they were grown under alternating freezing and thawing conditions. Meteorological records establish that an atypical alternating pattern of freezing and thawing conditions occurred in the spring of 1944, preceding one of the major outbreaks. See A. Z. Joffe, in *Microbial Toxins. Volume VII: Algal and Fungal Toxins*, edited by S. Kadis, A. Ciegler and S. J. Ajl (New York: Academic Press, 1971), 139–189.

19. *Global Warming and the Built Environment*, edited by R. Samuels and D. K. Prasad (London: E & FN Spon, 1994) is a useful resource on the relationship between climate change and the indoor environment.

20. Davis (2002).

21. C. M. van Herk, A. Aproot, and H. F. van Dobben, *Lichenologist* 34, 141–154 (2002).

22. A. J. Barsky and J. F. Borus offer a general review of the relationship between psychosocial factors and the amplification of symptoms in *Annals of Internal Medicine* 130, 910–921 (1999).

23. Kuhn and Ghannoum (2003).

24. J. Weiss, in *Proceedings of the 2nd NSF International Conference on Indoor Ait Health: Tends and Advances in Risk Assessment and Management* (Ann Arbor, Mich.: NSF International, 2001).

CHAPTER 8. A Plague upon Your House

1. M. Kettmann, *The Independent* (July 18, 2002).

2. M. C. Cooke, in *A Plain and Easy Account of the British Fungi* (London: Robert Hardwicke, 1871), wrote, "The *Merulius lacrymans* (*lacrymo*, Lat., I weep) is often dripping with moisture, as if weeping in regret for the havoc it has made."

3. To complete this exercise I must deconstruct *Meruliporia incrassata*. The *Meruli* part has been covered; *poria* refers to the porous nature of the fruiting bodies, and *incrassata* refers to their thickness.

4. See http://www.aquarestoration.com.

5. D. Arora, *Mushrooms Demystified: A Comprehensive Guide to the Fleshy Fungi*, 2nd ed. (Berkeley: Ten Speed Press, 1996).

6. Strands and rhizomorphs are related to one another in a developmental sense, because a rhizomorph develops from a strand. The term *cord* is used to denote an assemblage of hyphae that is intermediate in differentiation between the relatively loose bundle of cells in a strand and the fully integrated rhizomorph that grows only at its tip. *Serpula* forms strands but doesn't seem to produce rhizomorphs. *Meruliporia* generates both kinds of structure.

7. E. Kauhanen et al., *Environment International* 28, 153–157 (2002).

8. Eccentric mycologist Charles McIlvaine didn't try dry rot crusts, but he did taste related fungi with flattened—or "resupinate"— fruiting bodies, and wrote, "I have tasted, raw, every species I have found. They are all more or less woody in flavor, and I believe them to be edible." He concluded that these should be eaten only in an emergency, though he didn't specify the nature of the emergency. While we're on the topic, he also tried a number of fleshier bracket-forming fungi. Here are three of his reviews: "undoubtedly tough, but cut fine and stewed slowly for half an hour it is quite as tender as the muscle of an oyster and has a pleasant flavor"; "pleas-

antly crisp when stewed," and "edible if chopped fine and very well cooked." Among all of the bracket-forming fungi, he reserved scorn for *Polyporus heteroclitus*, saying, "As it ages it becomes offensive" (which is, in all probability, equally applicable to me). The quotations come from C. McIlvaine, *One Thousand American Fungi* (Indianapolis: Bowen-Merill, 1900). Mycologist David Arora (chapter 8, note 5) describes the dry rot fungus as "utterly and indisputably inedible."

9. Leviticus 14:33–53. "Fretting" refers to a gradual or insidious destruction, and "leprosy" is applied loosely here to imply a disease.

10. Closing up the home would promote fungal growth by restricting airflow and elevating indoor humidity. In other words, this action would help the priest appraise the problem. I'm being uncharacteristically judicious here in guessing that the priests were well meaning. (See contemporary cases of house cooking in chapter 1.)

11. Leviticus 14:50.

12. K. Bagchee, *Sydowia* 8, 80–85 (1954).

13. J. Singh et al., *The Mycologist* 7, 124–130 (1993). The expedition to the Himalayas found a single *Serpula* fruiting body in 1992; three previous searches had been unsuccessful. William Bridge Cooke found the fungus—which he called *Serpula americana*—during forays on Mount Shasta in California: *Sydowia* 9, 94–215 (1955). The presence in California of the fungus that causes dry rot in Europe is intriguing because *Meruliporia* rather than *Serpula* is responsible for dry rot in the United States.

14. J. Bech-Andersen, in *From Ethnomycology to Fungal Biotechnology: Exploiting Fungi from Natural Resources for Novel Products*, edited by J. Singh and K. R. Aneja (New York: Kluwer Academic/Plenum, 1999), 279–286. In this book chapter, Jørgen Bech-Andersen claims to have confirmed Cooke's discovery of *Serpula lacrymans* on Mount Shasta, when he visited the area in 1994. The website for the Hussvamp [house-mushroom] Laboratoriet in Denmark is another excellent resource for information on the natural occurrence of dry rot: http://www.hussvamp-lab.dk.

15. N. A. White et al., *Mycological Research* 101, 580–584 (1997).

16. S. Pepys, *Memoires relating to the state of the Royal Navy of England, for ten years, determin'd December 1688* (London: Ben. Griffin, 1690). For a more modern edition, see S. Pepys, *Memoires of the Royal Navy, 1679–1688* (New York: Haskell House, 1971).

17. B. Ridout, *Timber Decay in Buildings. The Conservation Approach to Treatment* (London: E. & F. N. Spon, 2000).

18. R. G. Albion, *Forests and Sea Power* (Cambridge, Mass.: Harvard University Press, 1926).

19. Similar contemporary use of unseasoned "green" or already infected wood for home construction has been implicated in the increase in wood rot in California caused by the fungus *Gloeophyllum trabeum*. Little attention has been paid to this fungus in comparison with *Meruliporia*, even though it may be responsible for substantial damage.

20. The loss of Henry VII's flagship, the *Mary Rose*, off Portsmouth in 1545 was probably due to the same cause.

21. John Ramsbottom furnished lively descriptions of dry rot in the British navy in the *Essex Naturalist* 25, 231–267 (1937) and in his classic book *Mushrooms and Toadstools: A Study of the Activities of Fungi* (London: Collins, 1953). A number of books on dry rot and the depletion of timber reserves were published in the nineteenth century. Robert McWilliam's *An Essay on the Origin and Operation of the Dry Rot, with a View to Its Prevention or Cure* (London: Taylor, 1818) gives a detailed view of the state of knowledge at this time. In another illustration of the ubiquity of fungal decay, I note that the pages of my copy of this 200-year-old book have been foxed by a black mold. Fungal foxing is discussed by Hideo Arai in *International Biodeterioration and Biodegradation* 46, 181–188 (2000).

22. Ramsbottom (chapter 8, note 21) cited a cost of £287,837 for the repair of the *Queen Charlotte* before she could be launched, and noted that after refitting "her name was changed to *Excellent*—a whimsical choice." In his book *Slayers, Saviors, Servants and Sex: An Exposé of Kingdom Fungi* (New York: Springer, 2001), mycologist David Moore estimated that in today's prices the repairs would cost $2.5 billion.

23. W. Blackstone, *Commentaries on the Laws of England* (Oxford: Clarendon Press, 1765).

24. M. Faraday, *On the Prevention of Dry Rot in Timber* (London: John Weale, 1836).

25. Kyan sold rights to the method to the Anti Dry Rot Company in 1836.

26. Faraday (1836).

27. The use of creosote as a wood preservative was patented in 1838 by John Bethell.

28. A fungus called *Lentinus lepideus* shows high tolerance for creosote and attacks railroad ties and telephone poles that are not impregnated with a sufficient quantity of water-repelling creosote at the time of treatment. Its common name is "the train wrecker." The fungus is probably implanted in the softwood timber before treatment, and forms its fruiting bodies when the moisture content of the wood increases through contact with soil.

29. Following an agreement with the wood-preservative industry in 2003, the EPA banned the use of chromated copper arsenate after 2003, but did not extend the ban to creosote.

30. M. Benjamin et al., *U.S. News and World Report* (November 6, 2000).

31. Unpublished opinion: *Joseph Glaviano et al. v. Allstate Insurance Co.*, No. 00-56754 (9th Cir.; 2002 U.S. App. LEXIS 9324). The court dismissed a second claim that the insurers had acted in bad faith when they denied insurance coverage to the family.

32. I. M. O'Brien et al., *Clinical Allergy* 8, 535–542 (1978).

33. E. A. Poe, "The Fall of the House of Usher," In *The Fall of the House of Usher: And Other Tales* (New York: Penguin, 1998).

34. C. T. Ingold, *Transactions of the British Mycological Society* 58, 179–195 (1978).

35. Bird's nest fungi do precisely the same damage, though their spore masses are propelled by rain drops splashing into the cup-shaped fruiting bodies. These are coprophilous fungi, too, and their range comes close to the performance of the artillery fungus.

36. See http://www.aerotechlabs.com.

37. Automobiles used to serve as an excellent food source for fungi. Wood-framed cars exported from the United States and Europe to the tropics in the first half of the twentieth century showed serious deterioration within months of arrival: See C. J. Humphrey, *Philippine Journal of Science* 46, 189–196 (1931). In his report from Manila, Humphrey wrote, "For cars in service longer than two years it is safe to assume that decay is at least well started. . . . It impresses itself upon the attention when the sills and vertical members have become sufficiently decayed to permit the doors to sag and be thrown out of alignment." The horror!

38. The fungus was illustrated in Sowerby's report to the navy in 1812, whose text was reproduced by J. Ramsbottom, *Essex Naturalist* 25, 231–267 (1937). *Sphaerobolus* also appeared in J. Sowerby, *Coloured Figures of English Fungi or Mushrooms* (London: Richard Taylor & Co., and R. Meredith; published in 5 parts between 1809–1815).

39. H. Fancher and D. Peoples, *Blade Runner* [screenplay], directed by Ridley Scott (1982).

40. Y. Karash, http://www.space.com (July 27, 2000).

Appendix: Mold Resources on the Web

National and local companies specializing in the evaluation and treatment of mold problems can be located with minimal effort on the internet or by consulting a phone directory. The author does not endorse any specific company.

A number of government agencies offer useful information for homeowners and for those who may have been adversely affected by exposure to indoor molds. The **U.S. Environmental Protection Agency's** Indoor Air Quality (IAQ) web site is very good:

http://www.epa.gov/iaq/molds

Useful documents titled "A Brief Guide to Mold, Moisture, and Your Home," and "Mold Remediation in Schools and Commercial Buildings" can be downloaded from this site, and links are provided to a variety of other resources, including information on asthma, on repairing a flooded home (from the American Red Cross), and cleaning carpeting (from the Carpet and Rug Institute).

The position statement on indoor mold from the **Centers for Disease Control and Prevention** (CDC) is available at:

http://www.cdc.gov/nceh/airpollution/mold

The U.S. Department of Labor, **Occupational Safety and Health Administration** web site provides other useful information:

http://www.osha.gov/SLTC/molds/index.html

Health departments in most states furnish their own guidelines about indoor mold problems. The **California Department of Health Services** provides a particularly rich mine of information on its indoor air quality web site:

http://www.cal-iaq.org

A number of private organizations offer valuable information for atopics and their families. The **American Academy of Allergy, Asthma & Immunology** (AAAAI) based in Milwaukee, Wisconsin, has an excellent web site:

http://www.aaaai.org

Their phone number for general inquiries is (414) 272-6071, and patient information and physician referral is available at 1-800-822-2762.

The **American Lung Association** web site is also helpful:

http://www.lungusa.org

In the book I mentioned **HarrisMartin Publishing**. Their web site represents the premier archive of information on mold-related lawsuits. Access to most portions of the site is available only to paid subscribers:

http://www.harrismartin.com

Insurance coverage for mold damage is detailed on the web sites of major insurers, and the **American Insurance Association** dedicates part of its web site to the mold issue:

http://www.aiadc.org/IndustryIssues/Mold.asp

Finally, no web-based investigation of molds is complete without a trip to the web site of Melinda Ballard's **Policyholders of America:**

https://www.policyholdersofamerica.org

Glossary

aflatoxin: potentially carcinogenic toxin produced by two species of the mold *Aspergillus*.

alimentary toxic aleukia (ATA): hemorrhagic illness thought to be caused by ingestion of grains contaminated by toxin-producing species of *Fusarium*.

allergic fungal sinusitis (AFS): a form of chronic nasal congestion in which fungi are implicated; also known as eosinophilic fungal rhino-sinusitis.

antigen: a molecule—or, more precisely, part of a molecule—recognized by an antibody.

apoptosis: genetically programmed cell death.

archebacteria (or archaea): one of the two major groups of bacteria, once thought to be restricted to extreme environments, now recognized as widespread microbes.

aspergillosis: human infection caused by a species of *Aspergillus*.

atopic: someone who suffers from allergies.

B cell (B lymphocyte): one of the two major types of lymphocyte.

choanoflagellate: microscopic aquatic organisms that are thought to be related to the common ancestor of animals and fungi.

chytrid: simple type of fungus whose cells swim by undulating a single flagellum or tail.

conidiophore: technical term for stalk that elevates spore-producing cells.

conidium (*pl.* conidia): type of fungal spore produced without sex between consenting partners.

dematiaceous: black pigmented.

dendritic cell: starfish-shaped cells that present antigens to T cells.

eubacteria: one of the two major groups of bacteria that encompasses all of the most familiar species, including those that live in our guts.

eukaryote: organisms consisting of one or more cells that house their chromosomes in nuclei (includes all of the fungi, plants, animals, and protists).

femtogram: 1×10^{-15} gram (0.000000000000001 gram).

fruiting body: mushroom or other multicellular structure produced by a fungus that contains myriad spore-producing cells.

hemosiderosis: accumulation of iron—from hemoglobin in red blood cells—within lung macrophages.

heterotroph: organism that meets its nutritional needs by digesting biological molecules assembled by other organisms.

hypersensitivity pneumonitis (extrinsic allergic alveolitis): farmer's lung and other inflammatory illnesses involving IgG.

hypha (*pl.* hyphae): filamentous cell produced by fungi.

IAQ: indoor air quality.

IgE (immunoglobulin E): class of immunoglobulins involved in allergic reactions.

IgG (immunoglobulin G or gamma globulin): the most common class of immunoglobulin in the bloodstream and lymph.

immunoglobulin: type of protein that includes antibodies.

IPH (idiopathic pulmonary hemorrhage): lung bleeding without apparent cause.

lymphocyte: major class of white blood cells whose surfaces bear receptors for antigens.

macrocyclic trichothecene: complex and highly toxic type of trichothecene produced by *Stachybotrys* (e.g., satratoxin G).

macrophage: large cells derived from the bone marrow that carry out a variety of crucial roles in immune responses.

mast cell: cells that bind IgE, degranulate (releasing histamine and other compounds), and produce a hypersensitivity reaction.

Glossary

melanized: darkly pigmented by the accumulation of melanin.

microfibrils: tough strands of molecules that form part of the structure of cell walls in plants and fungi, such as cellulose microfibrils (in plants), and chitin and beta-glucan microfibrils in fungi.

mildew: general term used to describe the appearance of microscopic fungal growth on a damp surface (see *mold*).

mold: general term for any microscopic fungus.

mycelium: feeding structure produced by a fungus, consisting of a network of branched hyphae.

mycosis (*pl.* mycoses): fungal infection of an animal.

mycotoxin: toxic metabolite produced by the mycelium of a fungus.

nanogram: 1×10^{-9} gram (0.000000001 gram, or 1 billionth of 1 gram).

picogram: 1×10^{-12} gram (0.000000000001 gram).

phialide: type of spore-producing cell that generates a succession of conidia.

prokaryotes: organisms whose cells lack nuclei (includes the eubacteria and archebacteria).

protease (proteinase): an enzyme that catalyzes the breakdown of proteins.

pulmonary hemorrhage: lung bleeding.

rhizomorph: root-like structure produced by a wood-decay fungus.

SIDS: sudden infant death syndrome.

spirocyclic drimane: a type of mycotoxin that upsets the immune system.

spore: microscopic fungal structure that functions like the seed of a plant.

stachybotryotoxicosis: poisoning caused by exposure to toxic metabolites produced by *Stachybotrys*.

stachylysin: hemolytic (blood cell bursting) protein produced by *Stachybotrys*.

T cell (T lymphocyte): one of the two major types of lymphocyte.

trichothecene: toxic metabolite or mycotoxin produced by certain molds.

VOC: volatile organic compound.

Index

Index